Penguin Education

Penguin Modern Psyc
General Editor: B. M. Fo

Occupational and Organizational Psychology
Editor: Peter Warr

Man–Machine Systems
W. T. Singleton

W. T. Singleton

Man-Machine Systems

Penguin Education

Penguin Education
A Division of Penguin Books Ltd,
Harmondsworth, Middlesex, England
Penguin Books Inc, 7110 Ambassador Road,
Baltimore, Md 21207, USA.
Penguin Books Australia Ltd,
Ringwood, Victoria, Australia
Penguin Books Canada Ltd,
41 Steelcase Road West,
Markham, Ontario, Canada

First published 1974
Copyright © W. T. Singleton, 1974

Made and printed in Great Britain by
Cox & Wyman Ltd, London, Reading and Fakenham
Set in Intertype Times

Contents

Editorial Foreword

Recent advances in psychology are important in many walks of life. This book is one of a series in occupational and organizational psychology which aims to illustrate how these advances are significant to people at work. The series will cover quite different issues ranging from the social organization of work to questions of individual performance and abilities.

The present book examines the design of work from a systems viewpoint. The systems in question have men and machines as their components, and it is the integration of these components which provides the author's focus. He is concerned with the allocation of function between man and machine, emphasizing how this should be seen as a matter of delegation from the former to the latter. He then deals with the analysis of work-tasks, skills, and errors, moving on to questions of selection and training. The scene is thus set for an examination of the complex interfaces between operators and their machines.

This book is a most timely one. Developments in the field have been rapid over the past two decades, and earlier texts are soon superseded. Professor Singleton provides an up-to-date and comprehensive coverage. The fact that he can do this so well in a limited space is partly due to the liberal use of figures and tables; these are themselves notable features of the book. In addition, many of the publications cited in the References are in only limited circulation; we have here a distillation of knowledge which the reader cannot easily acquire from other publications.

A rather different source of the book's timeliness comes from the fact that it is itself positioned at an interface – that between psychology, engineering design and management. Readers who are professionals or students in any of these fields can expect to find in it material of practical interest and use. Each of these three disciplines has much to learn from the others; in part this book is dedicated to their growing collaboration.

PBW

Introduction

It would be reasonable to introduce a book on systems by explaining what the term means, but this is not so easy to do formally. One problem is that a definition requires that something is described in more fundamental terms, but there are very few concepts which are more fundamental than that of a system. The physicist has a similar difficulty when he tries to explain what he means by energy; he usually resorts to explaining energy in terms of power and then completes the tautological circle by explaining power in terms of energy. His problem is that energy is a quite fundamental idea, analogous to an axiom in mathematics; one just has to accept it and synthesize other ideas and explanations from it. Another problem is that the term 'system' has more than one meaning. According to the Oxford Dictionary it has three: firstly a set of connected things, for example, the solar system, a body of knowledge or a doctrine, secondly a religious system and an orderly arrangement, and thirdly a botanical classification system. Within this book the first of these three will be used.

If we grant also that the physicist has developed more fundamental ideas than the ones we are going to use we can try to explain what is meant by a system in terms of basic physical ideas such as time, energy and information. A system is a set of related objects. 'Related' implies that information and energy are exchanged or shared between them, this in turn implies that they change in time. To put it another way all important systems are dynamic; there are static systems but they do not come within the scope of our study, e.g. a brick wall is a system but not a very interesting one. An 'object' is a point of interference with energy, they are located and identified by the sense organs. For these to function, the object must interfere in some characteristic way with energy thereby generating information, for example when light is reflected.

It is tempting also to propose that a system must have an *objective*, that is a purpose or a reason for existence. This brings us to the first sub-division of systems: there are *natural systems* and there are *man-made systems*. Natural systems obviously do not have objectives; it is not profitable to discuss the reason for the existence of the solar system or the purpose of a tree or the objective of a river.

An *engineer* is a person who designs or constructs. Systems-engineering then, is concerned with man-made systems although we must add the proviso that these man-made systems are constructed from components which have their origins in natural systems. Sometimes indeed the natural systems are transferred intact into man-made systems. The most general example of this is the human operator, a natural system but a necessary component of almost all man-made systems. It is sometimes argued that systems-engineering is a new name for the traditional general engineer. This somewhat parochial view accepts that the tendency for engineers to become more specialized and therefore narrower may have gone too far. Systems-engineering is an attempt to reverse this trend and to re-emphasize that good design requires a broad but balanced approach with weighted consideration of the range of all possible alternatives to the solution of particular problems, with adequate attention to the ways in which the solution of one problem will interact with potential solutions of other design problems within the system. This is a parochial view, valid as far as it goes but it goes nowhere near far enough. Design can no longer be confined within the limits set by the traditional engineering professions.

The change from thinking about machine design to thinking about systems design is not a gradual evolution but a discontinuity; an abrupt change in basic philosophy comparable to the change which occurred when the engineer took over design problems from the blacksmith and carpenter. The original approach to making machines involved a personal experience of the task for which the machine was required and an intuitive knowledge of the relevant properties of materials – a 'feel' for wood and metal. The machine was constructed from the materials by a process of trial and error. As the ambitions of machine-makers increased it gradually became apparent that these skills were inadequate. A systematic numerical knowledge of materials and power sources emerged and techniques were developed for trying ideas without using the real materials, e.g. the scale drawing. The engineer had arrived. Notice that the arrival of the specialist created at least one undesirable tendency, the designer was no longer necessarily a user or even a maker; this created a potentiality for communication failure. The machine became an end in itself with the needs of the user and the requirements of the operator relegated to the role of tiresome unquantifiable constraints. The timing of this change was different in different industries, but the turn of the century is a reasonable average point. The era of pure engineering only lasted for about half a century. The hardware-centred approach was perhaps a necessary phase to provide the stimulus for the development of technology based on the physical sciences but it was only a temporary one. The change to systems-engineering ended the era almost as suddenly as it began. As machines have got more and more complex there has been a switch to this new approach which has three characteristics: a consideration of function independently of mechanisms, a detailed attention to objectives and a consideration of man as an integral part of the system. It is this last point, of course, which justified the involvement of the human scientists in problems of design.

To understand the emergence of systems-thinking it is useful to identify the factors which have made it possible and necessary (Table 1). The underlying cause is the successful development of engineering technology. The engineer has not only solved problems, he has in many cases provided a whole variety of ways of solving a given problem. This

Table 1 **Characteristics of systems design thinking**

Cause	Characteristics	Consequence
Development of varieties of ways of carrying out the same activity	Consideration of functions independently of mechanisms	Improved systematic choice between alternative solutions
Increased complexity of new systems. Feasibility of system performance prediction	Detailed attention to objectives	Fewer expensive design errors. Reduced frequency and magnitude of modifications
Hardware performance becoming comparable with, and superior to, human performance	Consideration of man as an integral part of the system	Improvement of system performance by improving human performance

aspect has given rise to the need for functional thinking. If the only engine available is a steam engine then there is no restriction in thinking about a design problem with a steam engine in mind as the prime mover. When, however, there are electric motors, air motors, hydraulic motors, petrol engines, diesel engines, etc., available, it increases versatility to do the initial thinking about a design problem in functional terms such as prime-movers and power sources. Too early a switch to thinking in physical terms such as electric motors may unnecessarily restrict the solution of the problem.

As technology has developed and the complexity of products has increased, the cost of both the design process and the product have increased correspondingly. One has to be increasingly careful in arriving at design decisions on what the objectives of the product are and on exactly how they are to be met. For example, in designing a simple dynamo, one can afford to be relatively vague about what the ultimate performance will be and about what kind of materials and sizes to use for which components, but in designing a nuclear power station this kind of waiting to see simply will not do; one has to know exactly what the objectives are as well as when and how they will be achieved. It is only possible to embark on a very ambitious design such as nuclear power stations when one can predict with confidence the performance of the product before it is produced.

Another consequence of successful technology is that the performance and needs of the operator become an important set of variables. When machines are primitive, no one worries about the operator, any discomfort is more than compensated for by the excitement of a new machine and the limits on the system performance are set by the mechanisms. Eventually the action of the mechanism gets faster or more reliable than the reaction of the operator; further improvements in system performance are now best achieved by studies of the operator rather than more studies of the mechanism. For example, in the early days of the motor car the problems are to do with hardware, the designer is struggling to develop an engine that does not fail every few miles and a suspension which does not break on every hole in the road. The one who has the greatest success in these matters sells the most motor cars. Eventually the technology does develop. Engines and suspensions all become highly reliable, there is nothing to choose between the various products from this point of view. The successful designer is now the man who switches his attention to the comfort of seats, the positions of controls, effective visibility and so on, in general the human operator problems. This is where there is now scope for improvement and the possibility of doing better than one's competitors. The same transition occurs in relation to safety aspects of cars. In the early days it is a problem of better brakes and more reliable steering mechanisms. When these problems have been successfully overcome, safety becomes a problem of human judgements and attitudes.

All these are internal factors within the professions associated with designing and making things. There are also external aspects of the designer himself as a human operator and the broader impact on the whole society now pervaded by technology.

Technology has gone through an explosive development phase in the last century. The initiation of this phase depended largely on one factor which is usually forgotten, the productivity of farmworkers. The primary condition of survival is adequate nutrition. Mankind has been held back because for most of human history a worker could only generate enough food to feed himself and his family. As the productivity of farm workers increased it became possible to release a few active people from the task of generating food and providing the related protection from the environment. A few of these people devoted their energies to the design and construction of machines. When this happened we embarked on a process of development with positive feedback. The machines increased the productivity of agriculture still further, freeing even more people to think about even more machines. The stage was then set for exponential growth. In this phase at least, it seemed that a man plus a more elaborate machine produces sufficiently more to reward the investor in the machine, to pay for it and still increase the rewards of the operator. There have to be other developments in parallel such as legal, fiscal and administrative systems, but all of these are possible with increased reservoirs of spare worker capacity. It seems that the more one spends the more one has to spend, the more knowledge advances the faster it advances. Eventually there has to be either disintegration or a levelling of the growth curves. These apparently exponential curves must turn out to be the early parts of sigmoid curves. In fact there are some signs that this is now happening. The underlying factor which is now causing some levelling-off seems to be the rate at which individuals either can change or are prepared to change. Within the technological world these strains have been apparent now for several decades. There is a limit on the rate at which an individual can learn and a ceiling on how much he can know. At the scientific level research workers have become more and more specialized to the point which is proving self-defeating. For example, too much psychology makes one become a specialist within a narrow part of psychology but then, it turns out that to make progress one needs to know not only psychology, but also physiology and physics and engineering. Thus the individual psychologist cannot concentrate on thinking productively about psychology because he does not have the relevant knowledge of other related disciplines. If he spends his time learning the related disciplines he has no time to think about psychology; whether he does one or the other he certainly does not have time to keep up to date with all the work of the other psychologists. Inevitably the rate of advancement of real knowledge decreases.

There is a corresponding braking factor on technology. For the past century not only has technology advanced, but the rate of advance has increased. Thus we are off in another vicious circle. As knowledge advances systems get more complex. As systems become more complex they take longer to produce, but although they take longer to produce they become obsolete faster. In weapon systems and in aeronautics this process reached the absurd stage about a decade ago when systems were obsolete before they became operational. We are probably already through this phase. The VC10 will have a longer operational life than the Britannia basically because although people (airline passengers) were prepared to pay for the marginal improvement of the 1960s airliner over the 1950s airliner, they are not prepared to pay the much larger cost of the 1970s airliner over the 1960s airliner.

Nevertheless the designers are still trying to cope. The process was foreseen twenty years ago and it became very serious in the field of weapon systems (Singleton, 1971a). So much so that it became apparent that success in competition between nations depended on the development time for new systems. In an era of rapid technical change the competitor with the shortest development time is bound to win. The principle applies not only in military situations but also in industrial situations and it was the basic stimulus for the development of systems techniques. The broad aim is the faster development of better systems.

It may seem odd, at first sight, that there is no discussion of motivation within this book. Given that systems principles depend on the concept of objectives or goals and it is recognized that these are set by people and that systems are steered towards them by people, it might seem reasonable to incorporate some discussion of the drive towards these goals; in other words to discuss morale, motivation and incentives. This is not included because this book is about man–machine interaction. To follow the argument that such design problems cannot be considered without looking at needs and drives would be to invite the corresponding arguments from the other end of the spectrum; that design problems cannot be considered without looking at engineering problems of mechanisms.

The systems approach is used in the context of human motivation. There are a variety of different approaches which are based on systems concepts (Emery, 1969). Man–machine systems philosophy has some common factors with other systems philosophies, notably the emphasis on objectives and the importance of interaction effects or context, but the unique characteristic is the emphasis on man–machine interaction. For this reason also there is no discussion of interactions between people. This is obviously involved in all kinds of system problems: designs are done by teams, every machine operator has a supervisor, there are many machines that require more than one operator and there are many systems which involve complex communication between operators as well as between men and machines. However the man–man interface has different problems and different principles from the man–machine interface. These former aspects have been discussed in Brown (1954), Pym (1968) and Warr (1971).

In common with other system approaches, this book is not

really about how to deal with a problem but rather how to separate, identify and approach problems. Even the middle section of the book, which is concerned with analysis, contains description only at the 'what to do' rather than 'how to do it' level. There are two reasons for this. Firstly systems theory is essentially relevant to the structuring of problems rather than their solution. Secondly there are no formal procedures which can be described in detailed sequence which will solve a problem involving human behaviour. Any method, whether it involves observation, interview, survey or experiment must have a high content of creativity applied by the investigator to the particular problem in hand. There are of course detailed standard procedures (Singleton, 1971b) but these are different from systems methodology which supplies a context for the procedure and they are never recipes which, if followed, will guarantee the right answer.

Having said what the book is not about it would be appropriate to conclude this section by repeating what it is about. It is to do with the design of work, on the assumption that work nowadays is never done by men, nor is it done by machines, it is always done by man–machine systems. The man–machine system has proved enormously successful because men and machines are so different, each compensates for the weaknesses of the other. There are therefore problems of deciding what men should do and what machines should do in the pursuit of any objective. This is what man–machine allocation of function is about. Again because they are so different there are inevitable problems of ensuring that men and machines communicate effectively. This is what the design of man–machine interfaces is about. The smooth efficient interaction of men and machines cannot be achieved entirely by design. Certain kinds of men are best for certain kinds of machine control and furthermore they need to learn in order to achieve the highest level of efficiency. This is what selection and training are about.

Human performance, as a designer or an operator, is so complex that specialized techniques are needed to identify and separate the various human functions. This is what systems methodology and analytical techniques are concerned with. The whole makes up an integrated package of human knowledge, developed almost entirely within the last twenty years, which ought to be known about by engineers, designers, psychologists and administrators. If they all knew something about it, they would at least be able to communicate with each other more successfully. They might thereby generate rather more mutual sympathy and perhaps even empathy.

1 The Systems Approach

Systems Methodology

The purpose of the systems approach is two-fold: to cope with increasingly complex systems designs and to reduce development times for new systems.

There are three aids to the solution of problems generated by sheer complexity: the clarification of objectives, the separation of functions from the means of achieving them and the structuring of the design process itself.

Objectives

The traditional directive to the design engineer is that he should do the best he can with a given sort of problem. Exactly what is the best attainable and what resources of time or money are needed to achieve it is evolved slowly by dialogue and iteration between the sponsor of the project and the designer. There are many products for which this is still the optimum design procedure but as technology has advanced there has been a parallel increase in rigour of approach from two points of view. On the one hand more definite objectives are necessary because of the size of effort required. The more time and money that are involved the greater the pressure is to know what will finally emerge. On the other hand, more definite directives are acceptable to the designer because of his confidence in his own technology together with the degrees of freedom which it provides.

Details of objectives in the form of specifications are usually best provided in terms of inputs, outputs, times and costs. The different kinds of inputs and outputs must be specified preferably in numerical terms although ranges or probabilistic distributions often represent the greatest precision attainable. For example, no sponsor would ask for a new aircraft design which will fly as fast as possible with as long a range as possible. What happens is that high-density routes are identified, and this determines the required range, and an even number of trips within a twenty-four hour cycle fixes the speed. It is obviously not as simple as this; there are large numbers of constraints but the ones mentioned illustrate how an idea is refined to a specification before it is given to the designer. To take a less complex example, no sponsor would ask for a 'good' domestic washing machine. He would first of all look at the size and rate of change of the market and at its current distribution between various competitors. He would then look at the cost of advertising required to break into a given section of the market at a given level and would thus arrive at a fairly precise specification. In general the objectives stage provides the framework and the stimulus for two kinds of parallel studies: technical feasibility and market requirements.

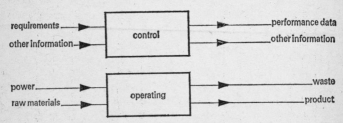

Figure 1 General double-block diagram of objectives

Specifications are best presented in the form of double-block diagrams as shown in Figure 1. Every system has a control part which deals essentially with information and an operating part which deals essentially with things, stuff and power. Some applications of this diagram to particular systems are shown in Figure 2. These are only at the very general level, one stage beyond Figure 1. It is necessary to go to further stages in order to identify and separate all the

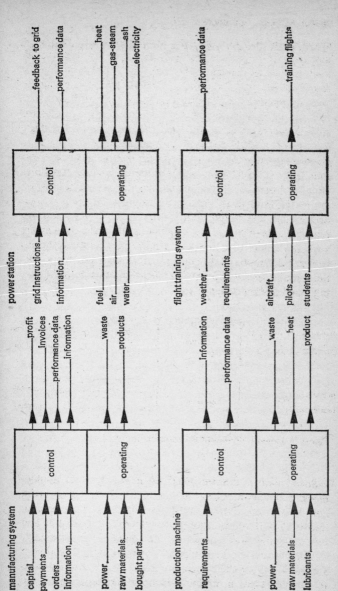

Figure 2 Examples of double-block diagrams

inputs and outputs and to describe their ranges of acceptability.

Functions

A function is an activity or set of activities described strictly in terms of activities and not in terms of means of achieving them. For example, power generation is a function but a petrol engine is not, it is a means of achieving that function. Unfortunately most designers, particularly engineers, are not taught to think in functional terms. Their education and training is mainly in terms of physical mechanisms and it is extremely difficult to persuade them to think at the more abstract level of functions; there is an element of truth in the old slander that the mechanical engineer cannot understand anything unless he can screw it in a vice and hit it with a hammer. Electrical engineers are somewhat better perhaps because electricity itself is an abstract concept rather than a physical entity. Of course there is another side to the argument and the engineer can often, with justice, point to the limitations of industrial designers and others who propose solutions to design problems which cannot possibly be realized in physical terms. Clearly there is a need for both and for an optimum timing of the switching between thinking in physical terms and thinking in functional terms. A start at the functional level is more and more essential because of the increasing variety of ways in which a particular problem can be solved physically. Pose a problem to a mechanical engineer and he will come up with a mechanical solution; pose the same problem to a hydraulics engineer and he will arrive at a feasible solution in hydraulic terms. To avoid this it is useful to think initially in functional terms thereby avoiding the pre-empting of the solution. Thus, in principle, the system designer has mechanical, hydraulic, pneumatic, electrical, electronic, chemical and electrochemical methods and mechanisms at his disposal. To choose between them and to provide a common language through which all the specialists can communicate we need functional type-thinking. Incidentally the whole range of types of mechanism are used within the human operator which is the ultimate example of complex but extraordinarily effective heterogeneous design. The opposite extreme of insensitive mixing of mechanisms is still too common. In the machine-tool industry for example there are many basically mechanical devices with a bit of hydraulics or electrics obviously tagged on as an afterthought. It ought to be possible to avoid and to eliminate this kind of clumsy design by generating functional studies on the basis of which various physical solutions can be evaluated and integrated systematically. The psychologist usually has no great difficulty about thinking functionally since this is his raison d'être. All psychological concepts which are not also physiological are essentially functional. For design problems the role of the human operator within the system is also best arrived at from a functional analysis of required system activities.

The Structure of the Design Process

Another inevitable consequence of the greater complexity of systems is that all separate design decisions become increasingly interactive and yet increasingly remote from the overall objectives of the system. Thus a structure for the design process itself is needed so that particular decision points can be perceived within the context of the whole process, and their consequences can also be more easily predicted. Perhaps most important, a structure will help to ensure that decisions are made in the optimum sequence or pattern. In order that such a structure should be general it must necessarily be phrased in rather global terms, but there are several levels at which this can be done. Perhaps the most general is that shown in Figure 3 where the whole design process for any system is summarized in just four blocks. This illustrates a number of fundamental points about design. For example, assuming that each block takes at least two years to achieve, it becomes clear why complex systems can take eight to ten years from initial conception to operational availability. The strategic thinking which led to the substitution of the process shown in Figure 3b for that in Figure 3a has been described

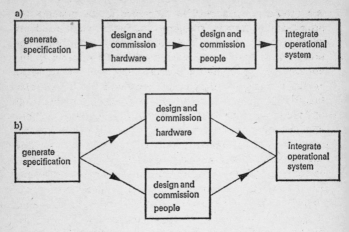

Figure 3 Reduction of development times

elsewhere (Singleton, 1971a). It is obvious that the process shown in Figure 3b is inherently faster than that shown in Figure 3a. It does, however, make greater demands on the designers. Firstly, it assumes that it is possible from a specification, to make some systematic decisions about the functions to be performed by men and the functions to be performed by machines. Secondly, it assumes that it is possible to create the personnel sub-system without having the final hardware available as a basis for selection and training. Thirdly it assumes that the separately created sub-systems, personnel and hardware, can be successfully integrated. To redescribe these activities in systems terminology the assumptions are respectively that we know about allocation of function between man and machine, about off-line selection and training and about man–machine interface design.

The description of what we do know about these areas is the core of this book. Note that they are all essentially human factors or ergonomic problems for which a knowledge of psychology is at least as important as a knowledge of engineering. These problems exist and always have existed for every design. The difference in the systems approach is that they are recognized as fundamental problems which require systematic solutions. The engineering approach in the past has been to regard them possibly as important but certainly as diffuse, not subject to rigid analytical methods and generally best left to others rather than to engineers. The traditional solution to the allocation of function problem is to do what can be done by machines and just leave the rest as a kind of literally indescribable remainder for the human operator. There is a case for this approach but it was stronger when hardware performance was so poor generally that the human operator could always be relied on to compensate for and to supplement the machine. It is an unfortunate fact of history that selection and training have developed independently of design problems. They are difficult enough for known machines and known tasks and it is only in very recent years that techniques have developed for task synthesis and extrapolation to skill descriptions. There have been associated developments in simulation, task training and team training so that it is now practicable to conceive of the development of the personnel sub-system in a concurrent rather than a sequential phase to the hardware development. Interface design has also developed beyond the stage of evolution by experience at least for the lower levels of skilled operator. Thus, just as a systems approach is not necessary until technology has developed from the physical sciences it is equally not feasible until other technologies of training and interface design have developed from the human sciences. Interface design is itself a particular aspect of total work-space design which also includes studies of environmental problems.

The total design process can be described as shown in Figure 4. It must be conceded at once that the methodology and techniques along the personnel line are nowhere near as precise as those along the hardware line, but the design decisions are at least of equal importance.

For all the diagrams in this chapter feedback loops have been omitted. They obviously exist since designing remains

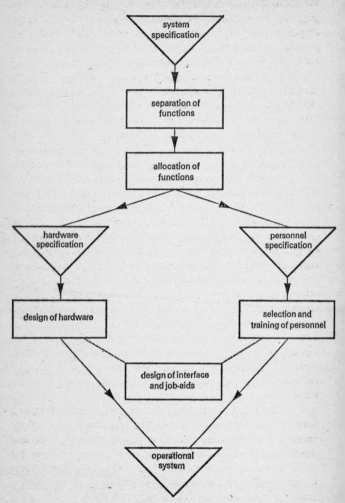

Figure 4 The systems design process

an iterative and not a serial process. The interactions take place at all stages and not simply through the whole process. One does not merely work through to the product and then reverse the specification in the light of user-experience. This remains the crucial final interaction but it is wasteful especially for large systems. Effective interaction takes place at all stages and all levels through the design process; this is one way in which the systems philosophy differs in degree rather than in kind from more traditional approaches.

One general difficulty about translating the systems philosophy into practice is in the identification of the boundaries of the system under consideration. The origin of this problem is in the inherent hierarchical nature of systems. Every system is made up of other systems and is itself a component of a larger system. It is too easy in designing one system to complain about the restraints imposed by the larger system; for example in designing a production system it is virtually standard to point out that the whole problem would be so much easier if only the marketing policy could be changed. It always seems easier to tackle the larger problem and the systems approach encourages this tendency. Sometimes it reveals valuable insights and at other times it merely confuses the issue. For example, in designing a motor-car it might in principle be helpful to consider that this is not necessarily the optimum means of transport, but more often, especially if one is employed by the design department of a car manufacturing company, such thoughts are a distracting irrelevance.

It is of some help to recognize that, as a design problem, every system is somewhere in the hierarchy between the simplest man–machine (one man plus elementary aids) and the whole world. In designing the simple man–machine, the problems of the world are only marginally involved, equally in the design of the world the problems of the simple man–machine are only marginally involved. For a given design a trio is under consideration, the system, its parent system and its sub-systems. In thinking about that system the designer must consider but must not get too involved in either the parent system or the sub-systems. There are of course designers for whom these sub-systems become the system and the system becomes the parent system. The zone of critical interest for a particular designer at a particular time must be kept in focus while recognizing that the total hierarchy exists.

Throughout this chapter most of the examples have been taken from man–machine systems where the design problems are clearly to do with mechanisms and tasks. Most people find this the easiest way to perceive the meaning of the systems approach. There are of course, other kinds of systems in which there may be little hardware and where procedures rather than men and machines are the dominant design problem. However, there is one set of problems common to all system designs, the problems of the people associated with that system, either as operators or users. Within the operators there are three categories: those primarily concerned with the setting-up aspects, the operating aspects and the maintenance aspects. Again it is misleading and over-simplifying to conceive of the human operator as concerned with on-line control. Automation removes this particular set of human functions but it does not remove the human factor problems of setting-up and maintenance.

The first two chapters are concerned with systems as systems involving men, machines and procedures. These separate categories are each important in their own right and warrant the attention of separate specialists. The remainder of the book is concerned mainly with the human factors aspects, the people problems of systems design.

Separation of Functions
Limitations of physical thinking

Functional thinking has been developed to compensate for the limitations of thinking in terms of physical entities. This latter is essentially conservative or convergent and it is also limited in range of comprehension.

The conservative aspect is best described by a number of examples which illustrate three different although related phenomena. Firstly the range of possible solutions is restricted to those encompassed by the particular kind of physical components. For example, one part of a car ignition system has, until recently, always been thought of as a mechanical distributor; a metal wiper rotating round a number of contacts corresponding to the number of cylinders. It requires a bit of functional thinking to escape from this concept and to consider electronic methods of generating a sequence of sparks at varying speeds. Secondly there is a tendency to innovate simply by change of scale rather than of method. For example, thinking about the principle of earth moving in terms of a spade shovelling inevitably leads to the design of a drag-line which is essentially a giant shovel pulled by wires and pulleys instead of by tendons and joints. One has to think functionally of general ideas such as the possibility of suspending the earth in water and pumping it away. Thirdly it is difficult, by physical thinking, to arrive at fundamental changes of method. For example, for half a century or more shoe-making machinery was designed on the principle of imitating the way the cobbler worked; this resulted in the most extraordinarily complicated mechanical devices for gripping a sole on to a last, putting tacks in the right position and hammering them home. Thinking in terms of the function of attaching a sole to an upper results in the development of appropriate adhesives and of welding techniques.

The problem of range of comprehension obviously becomes increasingly serious as machines get more complex. For example, it is possible to discover how a steam engine works by tracing the flow of heat, steam and water, and studying the relative timing of valves and the movements of the mechanical parts. This physical tracing technique will not work for a computer. The human brain cannot comprehend what is happening by looking simultaneously at all the different flows of current or potential changes in all the wires. The only way to understand what a computer is doing as a whole is to think in terms of inputs, outputs, arithmetic units, stores, data, programmes and so on. All of these are functional terms. Of course it is possible, having identified the functional units, to concentrate on one relatively very small part of the whole and then trace what is happening in electrical or mechanical terminology.

Both to facilitate comprehension and to stimulate flexibility of innovation we need to conceptualize systems at a more and more abstract or symbolic level. In particular it is necessary to separate the functional concept of a system from its physical realization; that is, to contemplate what it does as an issue quite separate from how it does it physically. Thus, one of the basic and early steps in the system design process is to separate out the functions required to achieve the overall objectives and only later to translate these functions into physical processes. The technique for both of these steps, identifying functions and switching from functional to physical units, is the block diagram. The power and universality of the block diagram has even led to the definition of the system designer as a man who draws block diagrams.

Advantages of block diagrams

The conceptual advantage of block diagrams is the ease with which the meaning of the separate blocks can shift up and down any scale of generality or complexity and across the conceptually difficult borderline between functional and physical entities. There are also very important practical advantages of the technique; it is possible to generate and display such diagrams on paper, blackboards, slides and so on.

The block diagram is an elegant technique which facilitates design because it has properties which match the needs of the human operator in thinking about a design problem. Returning to the systems-design process, the designer must form a concept of some system and this system, by definition, is an assembly of components related by a common purpose. Because of the limitations of his short-term memory and internal display system, the designer can only perceive a large scale system as a whole by condensing it into a small number of complex parts, and equally he cannot comprehend these parts without subdividing them into assemblies of even smaller parts. In other words, to think about a system of any size the designer must be able to shift rapidly up and down a hierarchy of building blocks. This technique avoids the limitations of his internal mental capacities already mentioned. One reasonable analogy of this process is the way in which a microscopist copes with the limitations of a fixed size viewing field. By reducing magnification he can shift from looking in detail at a small part of the total field to look at the total field in much less detail. For a particular purpose at a particular time he selects the appropriate compromise between the general broad view and the detailed narrow view. The analogy is not exact because the microscopist only shifts along one dimension-magnification, the block diagram provides several dimensions: physical–functional shifts, size shifts, complexity shifts, abstraction shifts and inversions.

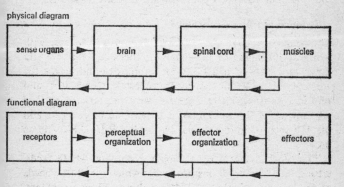

Figure 5 Four-block diagrams of the human operator

The problem of inversions is illustrated in Figure 5. The lower half of Figure 5 is a functional diagram which may appear to be the equivalent of the top half, but they are in fact very different.

Limitations of block diagrams

The limitations are mainly the inevitable byproducts of the advantages. The block diagram is a valuable communication device for members of a team from different disciplines, but because of its versatility it is difficult to arrive at an agreed diagram for a given system. Diagrams of the same system, drawn by two different people, rarely turn out to be identical unless they are very simple. It is common practice in examination papers on systems design to ask questions such as 'draw a block diagram of x'; x might be almost anything, a machine-tool, a post-office, a university department or a municipal bus system. Physical scientists and engineers are often disturbed to find that two answers to such a question can be marked equally high and yet be entirely different, nor is it possible to produce a model answer against which to mark other efforts.

The flexibility of block diagrams in switching from functional to physical representations is also a source of difficulty in that it is very easy to generate a diagram which is neither the one nor the other but a misleading mixture. This is a special case of the problem of level of abstraction. A diagram which is not at a consistent level of abstraction causes difficulties particularly at the allocation of function stage. Similarly the ease of shift up and down the level of complexity can facilitate the production of a diagram containing blocks at very different levels of complexity; this also is misleading. Some faults may be obvious but there can be much more subtle slips of this kind which are serious obstacles to design thinking until they are identified. In general a useful block diagram must have an element of unity, but the sheer variety of dimensions tempting the designer into disunity causes real problems. The most difficult faults to identify are those in which a diagram and its inversion have got confused.

The apparent similarity of physical and functional block diagrams of the same system can lead to the false assumption of identity of physical units and functional units. This problem is not so serious in hardware systems where, in fact, one criterion of good design is that the functional units are separately packaged in physical units, but this does not apply to biological systems. For example, visual perception is often discussed in the psychological literature. This is not a useful concept because the development of perception always involves more than one sensory mode. The error has arisen because of the attempt to identify a function (perception) with a physical data channel (eye-brain). An even grosser error is to identify the sequence of physical channel elements with functional elements by suggesting that the eye is concerned with sensation and the brain with perception. Finally, the flexibility of representation of the lines of a diagram can tempt the designer into generating hybrids where sometimes the line means that stuff passes, sometimes that data passes and sometimes merely that time has passed. An early mistake of this kind can cause members of a design group to be at cross purposes for a long time until the origin of the confusion is identified. All these errors may appear to be logically obvious and readily avoidable by tidy thinkers. Unfortunately it does not turn out in practice to be such a simple matter. For complex systems, even experienced design teams can get into considerable communication and conceptual difficulties by subtle and inadvertent errors of the kind discussed in this section. Very careful and time consuming effort and discussion of block diagrams are an essential investment in the early stages of any system design.

Display aspects of block diagrams

Designing displays is a problem of using the properties of some physical medium to match the structure of the information to be presented. This information is determined by the requirements and limitations of the observer. The observer in our case is a designer who wishes to form a concept of some system, and this system by definition is an assembly of components related by a common purpose. The human operator is accustomed to thinking in up to four dimensions

(when sticking close to reality he can conceive of one time
and three space dimensions) and it happens that there are
four dimensions available in a block diagram. Any graphical
display has up–down and left–right dimensions. The block
diagram also has the block–line dichotomy and each of these
can be treated as a dimension. For example the length or
style of a line can have meaning and so can the shape or size
of the block. In principle all of these dimensions can be
interchanged: a block-line diagram can be inverted to a path–
node diagram where the paths can represent what was previously in blocks and the lines become nodes. From this point
of view, designing a block diagram involves determining
what the designer requires and distributing this information
among the available dimensions.

There are a number of conventions or stereotypes which
restrict the total available choice. The decision on whether to
use a block–line or the equivalent path–node diagram
depends on the number of states to be displayed within each
dimension. A line or a path can be varied in length which is
an analogue presentation of state information and a block
can be varied in shape which is a symbolic distinction of state
information. Of course, a block could be varied in size but
this is undesirable because of the ambiguity as to
whether length or area is linearly related to the state being
displayed. In any graphical display, interchanging whatever
is represented from left to right with that represented up to
down, or vice versa, will not distort the information, although it may make it more difficult to perceive because of
display stereotypes. For example, from the reading and writing convention we are accustomed to left to right sequences
and thus time lapse is naturally represented in this way.
Up–down is ambiguous since reading and writing move up
to down but, on graphs, movement from down to up usually
denotes that something is increasing. Thus, whether the
primary input is on the upper or lower left and the primary
output is upper or lower right are arbitrary but the left–
rightness is not.

Generally, analogue presentation is better for continuous

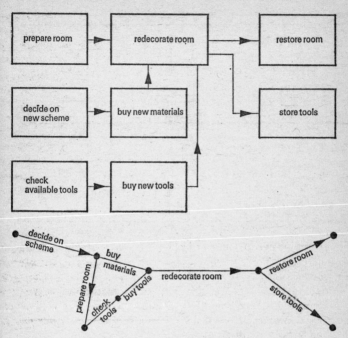

Figure 6 A block-line diagram and a path-node diagram for the redecoration of a room

information and symbolic presentation for discrete information. Thus, lines or paths are best for time, but there is a serious complication if feedback loops are included so that the identification of a sequence in time, such as a cause and effect, becomes impossible. Some accepted conventions for block shapes are shown in Figure 7.

Procedure for preparation of system diagrams

1 Define and list all inputs and outputs
2 Allocate the meaning of the available dimensions
   up–down
   left–right
   lines–blocks

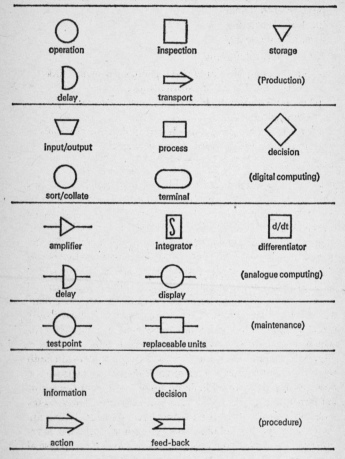

Figure 7 Some commonly accepted symbol conventions

3 Determine range and sequence of required diagrams
   physical–functional
   level of abstraction
   level of complexity
4 Avoid hybrid diagrams
   (a) lines should have only one meaning: sequence, time,
       energy, stuff, data or information
   (b) do not confuse dimensions and states within a
       dimension
   (c) blocks should have only one meaning (processes or
       things) unless different shapes are used

Conclusion

It is now possible to make a bald statement which would not
have been acceptable at the beginning of this chapter –
namely that separation of functions is a matter of drawing
the right set of block diagrams. What is meant by 'right'
depends on the particular kind of system, on the quality of the
design team and so on. There is no single universally acceptable solution. Nor is there a single appropriate way of generating the diagrams. For system designs which are a marginal
change from already operational systems it is easiest to start
by a physical analysis of the existing systems followed by a
translation into the functional equivalent. For systems for
which there is little previous experience it is better to start
with a functional diagram. It may be optimum to do both
separately and compare the results. Sometimes diagrams are
best generated by individuals or small teams and the teams
get larger as the detail of description increases. There seems
to be no substitute for or no equivalent to the block diagram
as a technique for defining and communicating what is required at all levels, from broad objectives to detailed physical
solutions.

2 Allocation of Function

Relative Performance Advantages of Men and Machines

The methodology of systems design appears to generate logically the problem of allocating functions. If we are to proceed from the stage at which required functions have been identified or separated and to translate the description of the design problem from functional to physical terms, we must allocate these functions between the available mechanisms. The special case of particular interest to us is the allocation of function between man and machine. In principle, the obvious way to do this is on the basis of the relative advantages of men and machines. The first attempt to define these relative advantages was made by Fitts (1951). The discussion of what are now called 'Fitts Lists' has proved to be a fascinating exercise throughout the past twenty years and every human engineering text-book contains a variation on this theme. One such list modified from Singleton (1966) is shown in Table 2. The list is essentially a rough, mostly qualitative balance-sheet of what men are good at and what machines are good for.

Clearly on matters of speed, power and consistency there is no comparison and no competition; the machine is invariably superior and, other things being equal, it should be used for functions in which these assets are of predominant importance. In the computer world it is now routine to talk in terms of nano-seconds (10^{-9} sec), whereas the human operator can only be expected to react in less than one second if the circumstances are entirely favourable, i.e. when he knows already more or less what the initiating stimulus will be, when it will occur and what response is optimal. The maximum power obtainable from a human operator, for example sprinting, is only about two horsepower and at this output rate the flow of blood to the muscles involved is inadequate to maintain performance for more than a matter of seconds. Even when working at a quarter of this rate he accumulates an oxygen debt which begins to restrict performance in a matter of minutes. The most obvious advantage of the machine is its consistency, the ability to repeat precisely and indefinitely a cycle of activity. The human operator is not reliable enough in this sense, there are always large random variations in, for example, his cycle time for any activity, and superimposed on this there are always also longer term trends due to learning and fatigue. Man is often described as a single-channel device from which it follows

Table 2 Fitts list – relative advantages of men and machines

Property	Machine	Man
Speed	Much superior	Lag one second
Power	Consistent at any level	2 horse-power for about ten seconds
	Large constant standard	0·5 horse-power for a few minutes
	forces and power available	0·2 horse-power for continuous work over a day
Consistency	Ideal for – routine repetition precision	Not reliable – should be monitored Subject to learning and fatigue
Complex activities	Multi-channel	Single channel Low information throughput
Memory	Best for literal reproduction and short term storage	Large store multiple access Better for principles and strategies
Reasoning	Good deductive Tedious to reprogramme	Good inductive Easy to reprogramme
Computation	Fast, accurate Poor at error correction	Slow Subject to error Good at error correction

Table 2 Fitts list – *continued*

Property	Machine	Man
Input	Some outside human senses, e.g. radioactivity	Wide range (10^{12}) and variety of stimuli dealt with by one unit, e.g. eye deals with relative location, movement and colour
	Insensitive to extraneous stimuli	Affected by heat, cold, noise and vibration
	Poor pattern detection	Good pattern detection
		Can detect very low signals
		Can detect signal in high noise levels
Overload reliability	Sudden breakdown	Graceful degradation
Intelligence	None	Can deal with unpredicted and unpredictable
	Incapable of goal switching or strategy switching without direction	Can anticipate
		Can adapt
Manipulative abilities	Specific	Great versatility and mobility

that he usually gets into trouble if he tries to do more than one thing at a time. The machine can be made to do more than one thing at once without difficulty, but this, of course, is a very superficial statement and comparison. It is not easy to define what is meant by 'one thing' and in any case since, for example, flying a helicopter is 'one thing', this is not a very restrictive property. In addition the operator has remarkable facilities for both task switching and for inserting what might be called a holding control circuit on one task while he switches his attention to another one so that he can fly a helicopter, fire a gun and communicate verbally apparently almost simultaneously.

In memory terms there is a contrast between the literal storage possible in machines and the rather vague store of ideas, principles and strategies with a relatively low factual content which the operator relies upon. On the other hand the operator does have the extraordinary and inimitable ability to access material apparently by a great variety of methods so that his equivalent of a cross-indexing facility is quite unique. For reasoning by deductive methods the machine is superior in that the computer can take over a human generated deductive system, but it will only accept such a system and function efficiently when the logic is impeccable. It is, of course, quite difficult or at least tedious to programme a machine. This is partly because the human operator does not, without rigid discipline, function entirely logically, he prefers to solve problems by what looks – on systematic analysis – to be a very untidy mixture of deductive and inductive processes. In performance terms he is much superior to the machine in that he can arrive at generalizations from badly structured evidence and he has not yet found effective ways of delegating this function to machines. By virtue of the sophistication of languages he is also very easy to programme. In simpler terms his goals, his methods and his attention can readily be redirected by written or verbal communication. For computational purposes again there is a superiority of machine performance over human performance. The machine is fast and accurate, the man is slow and subject to error; the saving grace of the human operator is his ability to correct errors. The basic human error rate is about 10^{-2}; any machine operating with such an error rate would be regarded as warranting repair. On the other hand when his error correction facilities are fully used the man can achieve an error rate as low as 10^{-6}, for example in landing an aircraft, and this puts him in a superior category to almost any machine. Human error will be discussed in greater detail in Chapter 7. Computation is not restricted to manipulation of numbers. For analogue processes also the machine is superior to the man. Thus analogue differentiation and integration to about 1 per cent accuracy can be done with cheap reliable mechanisms, e.g. electronic circuits, but human operators cannot easily achieve this level of accuracy. Compare for example human estimation of velocity with the performance of a speedometer.

With the exception of the special senses, machine data sensing is superior, for example in detection of all kinds of radiation. Even when specialized sensing is available the machine is relatively less sensitive to noise in the form of extraneous stimuli: heat, cold, noise and vibration. On the other hand there are no competitive hardware imitations of visual and auditory systems. Both of these have a vast range, about 10^{12} in energy units, and both operate along several dimensions. For example the eye as a sensor deals with relative location, movement and colour; for the auditory system the sensing and analysing mechanisms are even more mixed up and it is not easy to describe what the ear detects. Both eyes and ears can deal with very low intensity signals and the systems can detect signals in high noise levels. One example of this is the 'cocktail party phenomenon' in which it is possible to hear one's conversational partner regardless of the surrounding hubbub. This is related to the pattern perception facility for which the machine provides no comparable functions. Consider for example, the current difficulties in designing machines to recognize signatures or spoken telephone numbers.

On overload reliability, the difference between men and machines is nicely described by the American term 'graceful degradation'. Machine performance tends to degrade suddenly whereas men struggle on as loads increase and merely do progressively worse. For example Figure 8 shows schematically the behaviour of men and machines in a radar tracking task. At low levels of load, e.g. aircraft per unit of air space, the machine is much better than the man, but there comes a point at which the machine cannot discriminate the target aircraft from others and performance drops disastrously. The man does not have a corresponding point of discrimination failure – it just gets more and more difficult for him. The result is a region of high load (shown shaded) in which the function should be allocated to manual tracking rather than automatic tracking.

The real difference between human and machine performance comes from what we generally call intelligence. The machine has none and the man always has some. This manifests itself in his capacity to deal with events which the system-designer either did not predict at all – when a weapon system comes up against a previously unknown threat – or which were essentially unpredictable in detail, e.g. the traffic on a road. Man can also use his intelligence to compensate for his lesser abilities in other directions – he cannot react suddenly, but he can react quickly if he successfully anticipates what is going to happen. There are some functions which cannot be delegated to the machine not merely because technology is not sufficiently advanced but because such delegation might impede the basic man–machine relationship

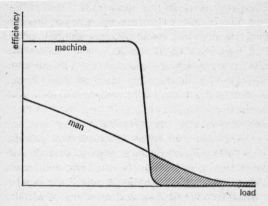

Figure 8 Tracking performance of men and machines

of the machine serving man. Thus the man must reserve the functions of goal-setting and goal-switching and sometimes of strategy switching, e.g. a human controller or pilot must decide when an aircraft landing should be switched to another airport.

The emphasis on higher level human functions sometimes results in the neglect of man's superiority to machines in manipulation. The motor abilities from locomotion to finger dexterity are not easily imitated by machines except in very restricted circumstances. For example, cars are better than men at moving along roads, but as the terrain gets more and more difficult there comes a point beyond which the man must rely on his own locomotor system. The study of design problems in prosthetic devices such as artificial arms generates a real awareness of and respect for the unique versatility of human limbs.

Differences between Men and Machines

The Fitts list approach was criticized fundamentally by Jordan (1963), who contended that the whole tenor of the argument was misdirected. The essential differences between men and machines, even for the problem of function allocation, are qualitative rather than quantitative. To attempt to quantify these differences is to obscure the issue and to disguise the fact that men and machines are essentially complementary rather than competitive. The basic advantages of men are in flexibility and adaptability; it is not possible to describe such properties by simple numbers. If the task requirements can be stated exactly then the machine will do a better job because it can be relied upon to be consistent, and thus it will best meet the stated requirements.

Following this criticism (the Jordan paper was available in 1961), Wulfeck and Zeitlin (1962) made a related criticism in technical rather than philosophical terms. They point out that principles of psychophysical scaling and the methodology of psychophysics are not strong enough to support the use of the Fitts list approach for specific design decisions. The problem is essentially a trade-off. What do you pay and what are you likely to get for your money? To answer this question there are obviously dominant economic factors but there are also technical psychological problems of obtaining and stating psychophysical and psychophysiological data in forms where such data can be used in the development of mathematical models of system performance. Most of the data obtained by psychologists are not in this form and, even more serious, most of the established methods are not capable of generating data in this form. Whitfield (1967) suggests that we need to take a broader view of human performance and to approach allocation of function through the study of human skill. Skilled behaviour is goal-directed and yet flexible in structure and strategy. These are the abilities we need and we must study them to improve our utilization of human performance in systems.

There are other distinctions between men and machines based, in materialistic terms, on the unity of the human operator. One can only incorporate human operators in systems as units; there are no halves or fractions and no separately functioning parts. Similarly incorporating two is never merely a matter of duplicating one. Good design requires not only that a given task should be within the capacities of the human operator but also that it make reasonable use of his abilities. It is bad design to expect him to have three arms, or eyes that look in two directions simultaneously. It is equally bad design to expect him to do nothing except press a button every half hour or to respond always with the same set of actions to the same set of immediate stimuli. This is called the problem of *the integrated task* and is another determinant of allocation of function. There are sometimes tasks or task elements which could be better (on other criteria) done by machine but they are allocated to the human operator to ensure that he has a job which is respectable. For example, although it is possible to incorporate a computer which will 'run up' a generator in a power station there is a case for leaving this as a manual operation because the operator must be there as a monitor and he may as well have the satisfaction and the increased arousal which results from direct personal control of the hardware.

Limitations of the General Approach

Extending the trade-off view of allocation of function, the cost–value ratios for human and hardware performance are bound to change as technology advances and they are also different for different countries or even for regions.

Allocation of function can be regarded as a functional location of the interface (Singleton, 1967a). Functions can be switched from human operators to hardware by, for example, improving information coding before display presentation or by manipulating the control dynamics so that less anticipation is required. Functions can be switched from hardware to human operators by increasing the complexity of human tasks and simultaneously increasing the quality of selection and training techniques. Thus allocation of function can sometimes be reduced to the question of what it will cost to pay more attention to the hardware in order to simplify the human tasks, as opposed to devising and using personnel techniques to ensure that the human skills are appropriate for the required tasks. There cannot be general answers to questions of this kind; answers depend on the relative availability of design skills and training skills and the kind of personnel available as operators. There are also, of course, various administrative and political variables such as what a contract actually requires, how close the relationships are between systems designers, contractors for hardware, training decisions and so on. These certainly cannot be treated generally; all we can do is note that decisions are often imposed by the nature of the decision system which itself exists in its present form for historical rather than logically defensible reasons.

Returning to the problem of the kind of personnel available, this is often much more relevant than any general statements about properties of man. It is the properties of the particular kinds of human operator which are going to be used in the system for external reasons which are important. For example, in military systems the allocation will change between countries where there is a system of conscription to one which relies on volunteers. The kinds and levels of skills of the ordinary serviceman will be different and so also will the optimum investment in training.

In industrial systems the problem becomes even more complex. Here it is often not so much what the men can do as what they will do within the context of the history of industrial relations and the culture of the region from which they are drawn. Even the factor of reliability changes dramatically and can have equally drastic effects on allocation of function. In a place of industrial strife the policy makers will often decide in favour of greater machine dominated systems simply because the machines are never deliberately awkward. However, this can have the reverse of the intended effect since there are always some human operators left and as their numbers decrease their individual power to disrupt the system performance increases. Thus the man part of the man–machine balance still has to be studied not so much by human performance techniques but rather by man-power techniques. The allocation of function must be based on the kinds of skills available in the relevant population.

The Central Role of Human Performance

The classical engineering approach to allocation of function between man and machine is to do what can be done at a reasonable cost by machine and to leave the operator to fill in the function gaps. Thus there was and is complete reliance on human versatility. For example, in most systems the human operator is the guarantor of reliability. His error detection and correction skills can be applied not only to his own performance but also to the machine performance. The design of a system which cannot rely on human availability for fault correction is of a totally different order of cost and complexity to that where the operator is at hand to deal with faults. This is one aspect of the predictability factor; fault prediction is impossible except in a statistical sense and thus the operator must deal with the specific cases as they arise. Another interesting aspect of predictability, or rather unpredictability, is that the presence of a human operator in a system means that the overall performance of the system is unpredictable. It also means that the system is adaptable in the context of unpredicted or previously unknown environments and in the face of unknown competitive systems. These two factors of inherent internal system unpredictability and the ability to cope with external unpredictabilities together account for the formidable nature of the man–machine system as a competitive device. This is expressed in the military context by the aphorism that the man–machine system will always defeat the automatic system.

It has been suggested earlier in this chapter that human performance is too important to be regarded as a reserve function and that there should be systematic allocation between man and machine. We might now go a stage further and suggest that this approach also does not adequately respect human performance. The problem is essentially not one of allocation of function but rather of delegation of function from man to machine (Singleton, 1967b). On these lines we can propose another definition of ergonomics: systems design with the characteristics of the human operator as the frame of reference. This puts the human operator unequivocally in the central role not only as the sponsor of systems but also as the key to the functioning of the system. There are many systems with more than one operator in them but nevertheless there is always one man who is responsible for achieving whatever purpose stimulated the design of the system. Thus, for a lathe the operator is the key operator, for an automatic lathe the setter is the key operator, for a factory the manager has this function and for an army the commanding officer is the key operator. There are many interdependent systems which can only be distinguished by identifying the key operator. For example, for the airliner as a flying machine, the pilot is the key operator, as a mobile restaurant the chief steward is the key operator and as a transport device the passenger is the key operator. Definition of these different identities through the key operator greatly facilitates precise thinking about the design of any given system and the related systems or sub-systems. From this point of view all hardware components and some human components are the extensions of the functions of the key operator (Table 3). Incidentally, as Craik pointed out in the 1940s (Sherwood, 1966), the development of technology can be traced through the efforts of man to extend the power

Table 3 Extensions of the key operator

Functions	Human extensions	Hardware extensions	Remaining problems
1 Manual	Slaves, servants, subordinates, teams	Tools axes $\rightarrow$ lathes	Dexterity (prosthetic devices) locomotion
2 Sensory	Scouts, subordinates	Time, direction, temperature, pressure, voltage, radiation	Pattern recognition
3 Decision making	Hierarchy of authority	Storage systems, feed-back systems, computers	Access to stored data, search and classification, inductive reasoning
4 Design	Draughtsman, clerk	Adaptive computers	Creativity

of his various functions. In order of development the first hardware extensions of human functions were those concerned with supplementing his muscle power; these are what we call tools and the phenomenon of man as a tool using animal has been used to distinguish man from other higher animals. Later he was successful in extending his sensory functions by spectacles, telescopes and compasses while he was also able to develop measuring devices to correspond to his concepts of the physical world such as clocks and thermometers. More recently he has greatly improved extensions of his computational function from the abacus to the digital computer. Extension of the human functions in the design and creativity area are still not well-developed.

The qualitatively different role of human operators has also been stressed by Bowen (1967). To use his graphic description the role of the human operator is that of the essentially adaptive element, a kind of elastic glue which functionally holds everything else together and enables it to work.

The presence of the man makes modifications easier. All systems are subject to design changes and the greater the degree of automation the more difficult it is either to identify required changes or to introduce them. This makes automated maintenance (see later) particularly difficult since a slight modification in the prime equipment may require a complete reappraisal of the automatic maintenance or fault finding system.

Historical Development (Table 4)

Sciences, and particularly human sciences, are more subject to the influence of fashion trends than scientists usually care to admit. That is, at any given time certain concepts dominate a particular problem area. This is not merely because they provide a timeless apposite model of the situation. The concepts are themselves influenced by current practical problems and also by current attitudes within society. The allocation of function concept, no doubt because of its central role in human performance studies and in design philosophy, is particularly prone to these influences.

The initial formulation in terms of a cold logical comparison of men and machines is understandable in the

Table 4 Historical development of the 'allocation of function' concept

Relative performance	Comparative assessment of human performance and hardware performance using the same indices
Cost–value ratios	The economic cost of hardware and human functions in relation respectively to their effectiveness in achieving system objectives
Integrated task	Task design to make full use of human advantages as well as to compensate for human limitations
Graded tasks	Provision of different levels of task to match individual differences and to provide promotion prospects
Function delegation	The use of hardware to supplement human functions which correspond to basic system functions
Flexible delegation	Provision of computer facilities so that the human operator can vary his degree of participation in on-line activity and can be assisted in system prediction

context of the immediate postwar world. The pioneer thinking in design philosophy was due to the designers of weapon systems. These men had acquired their training and experience during a long and ruthless war. It is not surprising that they were prepared to accept the problem as a simple one of relative performance of men and machines. This is not to suggest that Paul Fitts himself took such a simple view but rather that this single variable was seized upon from his and other writings.

During the early 1950s the cost of weapon systems escalated and it became obvious that the allocation of function problem should be considered in an economic context. In its simplest form the question became one of: 'is it cheaper to select, train and pay a man to do the task or is it cheaper to design, acquire and maintain a machine to do the same task?' It was and is never as simple as this, of course. The relative quality and reliability of output must be considered. The cost–value ratio appeared as the overall determinant of allocation. By this time also there had been some attempts to transfer the concept to non-military design problems. For jobs which are looked at as a life's work rather than for the period of a war it is evident that values are different, time spent in training can be an advantage rather than a disadvantage and job satisfaction becomes important. In this context it is not surprising that the concept of the integrated task emerged. This is essentially the design of a job which justifies using a human being rather than one which can be done by a human operator. When considering longer term jobs there are, of course, other problems outside the job itself. Prospects of promotion, possibilities of designing different kinds of jobs for different kinds of people are considered. This is the concept of graded tasks designed to provide for mobility between levels of job and to make optimum use of the wide range of abilities and capacities available.

Beyond this stage, allocation of function merges into the different problem of delegation of function. The human operator assumes the central role; this can be regarded as justified because of his unique performance capabilities or more basically because he is a human being. This takes us to the concept of the person with its implications not only of individual differences in performance but of individual differences in needs and interests. These factors can be incorporated within the design philosophy in many ways. It is now established practice in some industries to deliberately design man–machine tasks or ranges of tasks so that there are extensive differences in human involvement. At one extreme the job can be done without thought once the routine has been learned, at the other extreme continuous attention and decision making is required. Given sophisticated production control it is sometimes possible to allow the operator to choose the kind of task which he or she prefers either temporarily or on a more permanent basis. The difficulties of doing this and yet meeting customer requirements in terms of cost and delivery time should not be underestimated. Another solution possible in more complicated man–machine systems is to use the flexibility of advanced technology so that the machine takes over not only what the operator is not good at, but also what he does not feel like doing at the time. One solution to this design problem is called supervisory control.

Supervisory Control

This concept has arisen from the study of the design of teleoperators (Sheridan, 1972). These are, hardware mechanisms which provide power and manipulation facilities and are remotely controlled by human operators. Examples are human power augmentation systems and remote apparatus manipulators. One general design problem is that there are inevitable lags before the slave initiates and executes the commands of the controller. The problem is dealt with in two ways: the slave is provided with its own sensors and feedback loops so

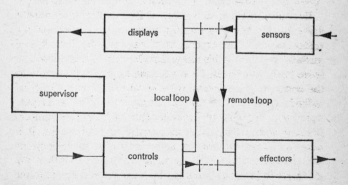

Figure 9 Supervisory control

that it can carry out routines with only general supervision. In addition a simulation system can be provided at the supervisor's end so that he can try out strategies or routines without waiting for the delays in trying them on-line and without the possibly disastrous consequences of on-line errors. The control system is shown in Figure 9.

In fact, this is a most elegant general solution to the allocation of function problem because the operator himself has flexibility of function allocation. He can allow the slave to function as an automaton and operate in a supervisory role or he can put himself in the control loops and function as a more orthodox man–machine system. Incidentally, the system provides excellent skill acquisition and skill maintenance facilities since the degree of supervisor control can be varied not only at the will of the supervisor, but also on a routine instructions basis or by an external system controller.

3 Total Activity Analysis

Task Analysis

A task is a piece of work to be done by the human operator. The Task Description is a key feature of any system design since it enumerates all the activities which are not carried out by hardware. In an operator-orientated design philosophy it is the basis from which the remainder of the design is carried out. Even in a system-orientated design it can form the criterion which initiates either a reallocation of functions or a change of system specification (Figure 10). It may be arrived at either by analysis or synthesis (Figure 11). The synthesis process results directly from the allocation of functions, the analysis involves the study of existing systems. It is desirable to be precise and consistent at least within a particular terminology. There is widespread confusion in the literature because different authors attach different meanings to what ought to be standard terms. For example, in many American publications the term *analysis* rather than *description* is used for the end-product of the process of study. It is also common to find *tasks, skills* and *jobs* hopelessly mixed up either as synonyms or with inconsistent shades of meaning. As shown in Table 5 there are two sets of terms: those describing activity levels in a rough hierarchy and those describing the stage, sequence or depth of the investigation. Any word from the first set of terms can be combined with any word from the second set to yield a term which some investigators have used at some time in work reported in the literature. There are twenty-four possible combinations and a much larger number of possible ways of defining each combination. Some are, of course, more common than others but attempts at standardization, e.g. DEP (1967), have not yet proved universally acceptable.

**Table 5 Terms used in the observation and
recording of people at work**

A (activity)	element	skill	task	job	occupation	role
B (stage)	study	description	analysis	specification		

Within this book, *task* is used in the work or system orientated sense and *job* with a human operator orientation. *Analysis* is used for the process of acquiring the evidence, *description* for the output of this process stated in overall terms and *specification* for the detailed functions required to achieve the descriptions. *Skills* are required within

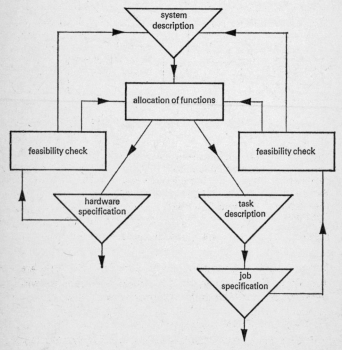

Figure 10 The iterative allocation of function process

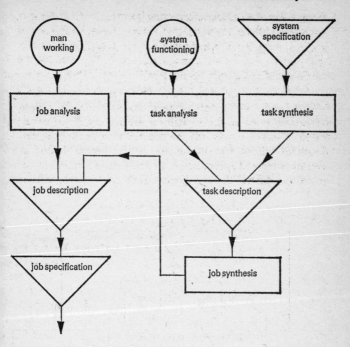

Task	system orientated
Job	operator orientated
Analysis	separating out the processes
Synthesis	putting together the processes
Description	the result of analysis in input–output terms
Specification	the result of analysis in functional terms

Figure 11 Descriptions and specifications for tasks and jobs

jobs and are the language of job specifications. *Elements* are discrete activities which can be defined as beginning and ending within real time as the task progresses.

The task description

It is desirable to have standard formats in which Task Descriptions can be expressed. One way of approaching this is from the point of view that the human operator is in a system essentially as a decision maker. Decision-making is choosing between alternative courses of action. A Task Description should then be complete if it enumerates all the possible courses of action together with the information required to choose between them. The human operator always has on-line and off-line inputs and his decisions depend on both. The on-line input conveys to him the current state of the system, the off-line inputs include such matters as policy, the probabilities of success of particular strategies and also the consequences which will follow given the success or the failure of a particular strategy (Figure 12). Sometimes, for

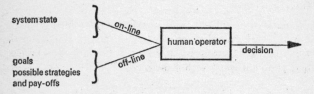

Figure 12 Factors in the task description

example, for a machine-operator, these off-line inputs are so standard and so well established that the problem reduces to that of defining the on-line inputs, but these cases are relatively rare. More often the off-line inputs are the key to the operator's behaviour, but because they are so varied and diffuse they are difficult to identify. For this and other reasons the Task Description is not usually complete and comprehensive. If it is, then we can evoke the Jordan principle (p. 44) and reconsider why the task has been allocated to the man. Thus, there is a paradox in that the Task Description is necessary to continue with further stages of the personnel design, but if it is possible then it probably should not be a personnel problem since the man could be replaced by a mechanism.

The description problem is partly determined by the analysis problem, partly by the objective and partly by the needs of research and scholarship. It is no good postulating that certain evidence is required if there are no conceivable means of acquiring it. Conversely it is no good just writing down that which is easily obtained without questioning its relevance to the objective. In general, the whole field is in a state of flux and there is little accepted knowledge and expertise. Thus, it is even more than usually important to work in cooperation with other colleagues with similar interests. That is, to generate evidence and conclusions which are comparable with those obtained by other investigators. This is a tall order and it is not surprising that progress is relatively slow and that all the techniques require considerable initiative and creativity on the part of the investigator.

Analysis techniques

1 Input–output models. The simplest comprehensive version of a task is that described by a single or double block diagram of the kind shown in Figure 1. In this case of course there is no hardware within the block. Figure 13 shows such

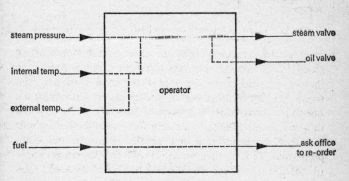

Figure 13 Task description for boiler operator

a diagram for a central heating system controller. Such an analysis involves identifying all the inputs and outputs and linking them together in functional chains. Usually this is difficult to do in practice. In a task which requires the operation of a control console following observation of the real world, for example operating a machine tool, the inputs are

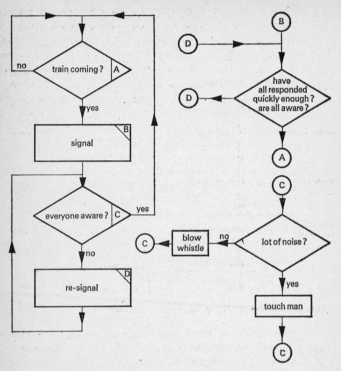

Figure 14 Task description for a railway lookout

difficult to identify. For a management task the inputs may be known (computer print-outs) but the outputs are difficult to identify. Generally the higher the level of the task the more difficult it is to specify unique input–output relationships.

2 Algorithms. This is a particularly useful technique when the operator is receiving a series of inputs and he makes a corresponding series of decisions. The problem is to identify the decisions and their order together with the relevant inputs. If possible the decisions should be reduced to binary

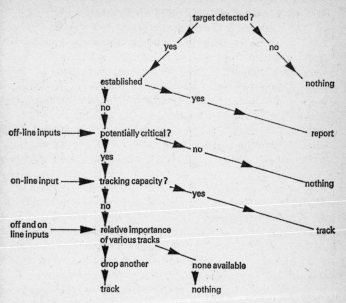

Figure 15 Task description for a radar operator

choices. Conventions about symbology are not very important providing the system used is clear and unambiguous,
Figures 14 and 15 show two examples using slightly different
conventions. The need to prepare logical sequences of these
kinds provides stimulus and direction for the investigator in
his utilization of observation, conversation, study of records,
etc. which provide the evidence needed to complete the sequence.

3 Link-design charts. Another way of structuring the attention of the investigator is to start from the machine and consider all the possible ways in which he can conceivably effect
its function. This is done by drawing a functional diagram of
the flow of power (Singleton, 1964). The next step is to identify how he executes the decisions he makes which effectively
means defining the relevant controls. It is thus possible to

identify what the controls do and, by observation of the particular tasks, to identify the sequence in which they are used.

The link-design chart itself assumes that for each control action there must be a trigger, often in the form of an action previously completed. The operator must have some check that he has done what he wanted to do; this is an information input which is compared with the purpose which initiated that particular action. Thus, if the sequence: purpose – trigger – action – check can be identified for each action within the task then the task has been described. These charts are particularly useful for tasks which are output-dominated.

4 Responsibility charts. If the task is input–dominated it is often possible to start by asking what the man is responsible for. Once these decisions and duties have been identified, the relevant questions which the operator must ask can usually be deduced. From these the information required to answer the questions can be studied and enumerated. This completes the task descriptions, since for this particular kind of task what he actually does about it is too varied to identify precisely. Table 6 illustrates this process.

Job Analysis

A job is a set of tasks, a task or perhaps part of a task which a human operator carries out because he has some motive for doing so. It would be useful if a complex variable such as motivation could be separated from job analysis but unfortunately this is not feasible. For example, it is accepted by those experienced in analytical work (Miller, 1967; Edwards, 1971; Annett, 1971), that what a man is doing does not make sense and cannot be described without reference to his purpose in doing it. Yet if one asks a man who, for example, is tightening a bolt, why he is doing it, he might legitimately reply with a range of very different answers: because he was told to, because he gets paid for it, or in order to help build a motor car. This example illustrates the need for precision in fixing the level of analysis and for restricting the description as much as possible to the problem in hand. In other words,

Table 6 Task Description for Production Executives

Executive	Decisions or duties	Questions to be answered	Information required
Director and General Manager	A To participate in controlling production system	(a) Is it running satisfactorily or not?	1 The amount of work at each stage 2 Whether any work has been at one stage for too long 3 Whether any lot of work is going to be late for delivery
Production Manager	B To regard particular orders as 'rush' rather than routine	(b) Is any lot of work likely to be late?	see 3
	C To participate in controlling any section of production system	(c) Is each section producing satisfactorily?	see 1
	D To convey information to Sales concerning position of any order	(d) What stage of production has a particular order reached?	4 The location of each customer's order within production system
Production Controller	E To convey information to ordering department concerning material requirements	(e) Is material available for any lot of work?	5 Materials position in relation to orders
	F To transfer work across the section boundaries	(f) Work in progress and delivery dates for each lot	6 5, 4, and 3, and the style and size range of each lot
	G To allocate work between the two factories	(g) What work is available for second factory?	7 Material number of each lot

what the investigator does is also determined by why he is doing it. His purpose may be to do with selection, training, work design or incentives and how he sets about it will be quite different for each of these objectives. This is not to suggest that there are a large number of available techniques but rather that the level of analysis and the interpretation of results are determined by the purpose of the investigation.

A further complication arises from the inherent system flexibility and adaptability provided by human operators. Inevitably the various people associated with a particular job have different frames of reference and different ideas about what the job should be. This is notably the worker himself, his immediate superior and the system key-operator (in the case of a production system this is usually the manager). This gives rise to at least four different answers to the question of what a particular worker is doing: what the system requires (the task), what the foreman considers should be done, what the worker says he does and what he actually does. These versions are often quite difficult to reconcile into an agreed job description, and again the different weighting to be assigned to the various sources of evidence depends on the purpose of the investigation.

All these factors combine to ensure that every job analysis is a specific activity about which there may be certain ground rules and a certain standardization of procedure but always with a high content of originality on the part of the investigator.

Job description

The variety of human effort does not encourage the possibility of a neat set of descriptors designed to cover everything that a man might do under the heading of work. Not surprisingly, progress was first made in the study of straightforward sequential tasks where the human operator is paid to go through a standard cycle of output activities, e.g. assembly tasks. All that is necessary in these cases are methods of describing the cycle clearly and unambiguously. The first complication occurs when there is no longer a standard sequence; the operator has a limited repertoire of actions he can take but the order in which he takes them depends on the flow of information, e.g. the process-control operator. It then becomes necessary, of course, to define not only his outputs but also his inputs.

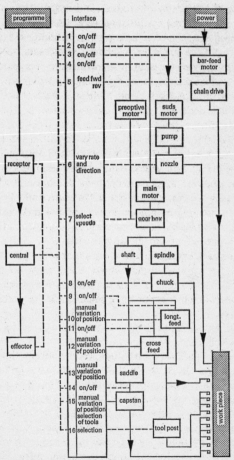

Figure 16 Capstan lathe

The next level of task occurs where there are no standard stimulus–response relationships and it then becomes necessary to define how the operator is making decisions, for example fault finding. At an even higher level the situation is known, the required answer is known but the best means of achieving it is not obvious. This occurs when the output is through other people rather than through a machine. What tends to happen in the analysis process is that the higher the level of task, the vaguer the description gets and the more the terminology changes. At the lowest level the actions are described, then the input/output relationship, then the decisions and finally the responsibilities. A particular job description will often be a mixture of descriptors at at least two of these levels.

Analysis techniques

If one wants to find out what a human operator is doing there are a limited number of ways in which it is possible to proceed even in principle (Table 7). One can observe him and note that which is observed, one can observe and infer what he must be doing, one can record the effects of what he does either on himself or on his surroundings, one can ask him and one can ask his peers.

Direct observation methods

The main problem here is that observers do not naturally function in a systematic or sequential manner. They have to be trained and disciplined to do this. The vehicles for training and discipline have been developed by the method study practitioners. They are all essentially charting techniques where a set of standard symbols is used to record what is happening. The records can be extraordinarily elaborate, e.g. within the *therblig* system there are specified symbols, specified kinds of shading, specified colours of pencils to use and even a specific time system (the wink $\equiv \frac{1}{2000}$ minute) (Shaw, 1952; Chapanis, 1959). The main point about these charts is that, in order to make a classified record of activity, the observer is forced to look carefully at what is happening

Table 7 Job Analysis Methods and Techniques

Method	Techniques
Direct observation	Process charts Micro-motion charts Flow charts
Observation and Inference	Sensori-motor charts Decision charts
Recording	Filming Instrumentation
Interview	Continuous commentary Critical incident Depth interview
Records	Depends on records

and to interpret the purpose of the action at any instant. They provide records which are sometimes useful and there are a few general rules about the use of the charts to devise new working methods, to reduce delays, aim for symmetry in two-handed process charts, etc. The main pay-off is simply that, having completed the chart, the observer has literally observed; he now knows more about the job and he can use this knowledge to suggest changes in the workspace, the interface, the procedure and the selection and training methods. This is the observer discipline aspect; the observer training aspect follows from practice in attempting to use the charts where even on very simple assembly jobs the attempt will lead to intricate discussion of exactly how to observe and classify observations.

1 Process charts. The best known classification system is that indicated by the American Society of Mechanical Engineers (ASME) symbols. These are astonishingly versatile not only in their comprehensive coverage of all human dexterous activities, but also in that they can be used, admittedly with slight variations in the meaning of symbols, to describe what the man is doing, what is happening to the material being worked on, or as an outline of a whole process. They can also be used at the more micro-level, again with slight changes in category definitions, to describe in detail what the hands are doing separately.

2 Micro-motion charts. The elements defined by symbols for these charts are much smaller, typically about a second or less. There are various classification systems such as the SIMO system which uses the therbligs already mentioned. There are about twenty of these with movements for example subdivided into search, select, transport loaded, release load and transport empty, and delays subdivided into unavoidable, avoidable and rest. The detail of the description is normally such that a film of the operation is required which can be subjected to frame by frame analysis.

The most popular systems now used are the Predetermined Motion Time Systems (PMTS) in which there is a classification of elements at even more detail than for therbligs. In addition there are standard times available for each standard element. One of the best known PMTS is Method–Time Measurement (Maynard *et al.*, 1956). It is possible to analyse a task, add up the element times thus obtained and either predict the task time or compare the times required using different procedures for a given task. There are theoretical difficulties in assuming that task elements are discrete and additive in this way, but although there may be a considerable error, such predicted times are better than no knowledge at all especially for tasks which are being synthesized rather than analysed.

3 Flow charts. These are normally pictorial diagrams which correspond topographically to the real situation, for example a large console or a machine shop, and lines are drawn to represent movements by the operator of the material or information. The line may be ink on paper, a piece of string (a string diagram) or the photographic trace of a light (a cyclograph). For the latter, the light can be made to flash regularly or the camera shutter can interrupt regularly thus providing a time measurement (a chronocyclograph).

Observation and inference methods

The direct observation methods are strictly only for human outputs. When it is necessary to analyse inputs and decisions as well as outputs a considerable element of inference enters into the charting.

1 Sensori-motor process charts. There have been various attempts to categorize human information processing stemming from a suggestion by Crossman (1956) that human control can be regarded as following the sequence: plan – initiate – control – end and check. He devised a symbol for each of these elements and added another column to the standard two handed process chart. In practice it usually turns out to be too complicated to make these detailed inferences particularly since the elements can overlap; the operator can plan his next move while he is still executing the previous one. This has led to various simplified versions, (Seymour, 1968). One is just to add another column to the two handed process chart and label it 'attention'. Another is to add three additional columns, one for the purpose, one for vision and one for the other senses.

2 Decision charts. These are the same as those used for Task Analysis with a difference of emphasis in that the man is the centre of attention and analysis rather than the system or sub-systems.

3 Robot simulation. A novel approach to the identification of jobs was suggested by Miller (1967). He proposed that one way of describing what is required to be done is to describe a robot which will do it. This leads to a Task Specification rather than a Task Description. Such a robot must have a purpose and to have a purpose implies long term and short term memory functions. Apart from storing, memory is needed for other factors relevant to decision making. To

OBJECTIVE Make king-pin
SYSTEM Capstan lathe and skilled operator
DATE September 1961 OBSERVER W.T.S.

No.	PURPOSE	TRIGGER Previous check / Next Action	NA	Control No. or Letter	ACTION Both hands / Right hand RH / Left hand LH / Foot LF/RF	BH	CHECK Vision / Hearing A / Touch Kinesthesis K / Taste Smell S	V
1	Provide power	Start		1	Up isolator lever		Pressure, position	
2	Preoptive motor on	✓		2	Press button		Red light	
3	Suds motor on	✓		3	Press button		Suds run	
4	Main motor on	✓		4	Press button		Noise	
				5	Capstan feed			
5	Position bore	✓		6	Release clutch		Position impact	
				5	Feed motor R F			
6	Provide first tool	✓		15	Index capstan		Pressure, position	
7	Adjust spindle speed	Experience		7	Speed control and knob			
8	Provide lubrication	✓		6	Adjust nozzle		Stream	
9	Apply first tool	✓		15	Bring capstan forward		Pressure, position, noise	
10	Apply powered feed			14	Lift control			
11	Adjust preoptive speed	Experience		7	Speed control		Reading	
12	Provide next tool	Powered feed thrown out		15	Index capstan		Pressure, noise	
13	Adjust spindle speed	✓		7	Preoptive knob		Appearance	
14	19 as 18-13	for 2nd tool						
20	-25 ditto	for 3rd tool						
26	-30 ditto	for 4th tool						
30	-35 ditto	for 5th tool						
36	--38 as 8-10							
39	Remove capstan head	Capstan cycle completed		15	Index capstan		Pressure, noise	
40	Provide 1st saddle stop		✓	A	Stop control		Position	
41	Provide correct post tool		✓	16	Tool post control		Position	
42	Bring tool into use	✓		12	Cross feed control		Position of tool	
				10	Longt feed control		with saddle against stop	
43	Provide lubrication	✓		6	Adjust nozzle		Stream	
44	Use 1st post tool	✓		10	Longt feed control		Pressure reading	
45	Withdraw tool post	✓		12	Cross feed control		Away from	
				10	Longt feed control		Work piece	
46	-51 as 40-45	For 2nd tool and stop						
52	Provide 3rd saddle stop		✓	A	Stop control		Position	
53	Position parting off tool	Tool post cycle completed		12	Cross feed control		Position of tool with saddle	
				10	Longt feed control		against stop	
54	Use parting off tool	✓		12	Cross feed control			
55	Remove work piece	✓			Catch piece and control		Appearance feel	
56	Remove parting off tool			12	Cross feed control		Away from	
				10	Longt feed control		centre line	

Figure 17 Task description for a capstan lathe operator

arrive at the decision, scanning, identification and interpretation functions are required, and to implement the decision effector responses are needed. This, however, is only one dimension of the situation.

The above categories only identify the functions. It is also necessary to describe the task content which copes with differences within and between individuals. For example, individuals differ in their strategies and in the 'internal model' they use to control and predict the dynamic situation. They also vary in their training and experience, which Miller calls the 'degree of learning' dimension. Thus, to describe the job a man is doing, we need to know about how the job is broken down into function categories, about the way he does it and about his degree of skill. Note that Miller is using the words 'task' and 'job' differently from the way they are used in this book.

4 General approach. The common-sense detailed study of all the aspects of a job which are relevant to personnel subsystem design is best summarized by Ferries (1967). He subdivides the job specification into reception of information, memory, interpretation and motor responses. The corresponding Task Description is obtained using the headings: function, information, simultaneity, critical errors, system emergencies and environmental conditions.

Recordings

The recording and measurement of performance has turned out to be rather a disappointing technique for job analyses, except at the global level of films which are just an aid to direct observation. It is usually easier to record outputs than inputs and even if both are available, the purpose of a particular output is not easy to identify from records because most human outputs are responses to a variety of inputs. In addition it is not usually possible to decide what to record until the job description is available, which rather puts the cart before the horse. The basic problem is that which keeps occurring in the definition of human work: if the definition is possible then why is a man rather than a machine doing the job? Recording becomes important in experimental rather than observation settings and it is thus most relevant to Skills Analysis.

Interviews

1 Continuous commentary. It is obviously useful to supplement the observation of what an operator is doing with his own comments on what he is doing. This can vary from occasionally asking him questions to asking him to describe all the time what he is doing and why he is doing it. This technique has been used as an aid to training, e.g. by advanced motoring schools, and it has been developed for job analysis of process operations such as paper-making and steel-making by Beishon (1967) and Bainbridge *et al.* (1968). It has proved particularly useful for process jobs where the operator is responding on a long time scale. That is, what he does at one instant is a function not only of what is happening now but what happened some time ago and what will happen some time in the future. Past and future in this context are measured in minutes rather than in seconds or hours.

2 Critical incident. It is characteristic of most real jobs that the level of work-load varies considerably. For most of the time the operator is working well below his capacity but there are infrequent occasions when his information input or output or energy output rises above or beyond his capacity. Clearly these occasions are important in understanding the job and one method of focusing on them is simply to ask the operator to describe occasions when he got into difficulties, or nearly did, or observed someone else in this kind of predicament. This technique is fundamentally one of error analysis and is described in more detail later.

3 Depth interview. Communication between people appears deceptively straightforward but to achieve a high level of such communication, particularly in an investigator–subject situation, requires great skill on the one hand and cooperation on the other. There are cultural and social problems as well as technical ones. For example the interviewer needs to be knowledgeable about the job and its context, he must know the jargon, the history, the taboos, etc. He must not be too much younger than the subject, his accent must not be obviously different, his attitude must be sympathetic and yet not effusive. The best interviewers have what appear to be neutral personalities thus minimizing their directing influence on the subjects. One good example of the use of this technique to investigate jobs is provided by Chadwick-Jones (1969).

Records

Every large system has various banks of data about itself and about the sub-systems. These are valuable sources of information about the jobs within the system. If the system was designed on system design principles then there should be complete records of all the task and job descriptions at least as they were when the system was first set up. The difficulty of course is that records are not necessarily kept up to date when jobs change and develop as they always do in any effective system.

Within a military organization there are usually manuals and drills which can be consulted. Within an industrial organization there are offices to do with production control, work study, wages, technical aspects and perhaps a central computer facility all of which contain data which can be valuable but which always need to be interpreted with caution. Broadly speaking the more remote the data bank from the operation the less valid the data will be.

Skills Analysis

This terminology is unfortunate in that what is analysed is often more than skills but it has the sanction of wide current usage (Seymour, 1968, Singer and Ramsden, 1969). Skills analysis is particularly orientated towards the training problem and therefore concentrates on the aspects of behaviour which are susceptible to improvement by learning. From this point of view the term 'skill' is apposite since by definition skills are learned and change with experience. This is one of three characteristics of skill as defined by Welford (1958). The others are that skills are essentially hierarchical – there are skills within skills – and are serial – many different processes and actions are ordered and coordinated in a temporal sequence. These latter characteristics inevitably complicate problems of analysis and description. From the serial nature of skills it follows that the techniques of separation into time segments cannot be used. For any skilled operation it is not profitable to identify that, at an instant in time, skill A stopped and skill B started. Skill and skill elements overlap, they come and go like themes within a musical score, and although it may be useful to identify when one or more than one is present or absent, any attempt to chop up the serial flow into discrete time intervals will destroy the essence of the skilled activity. Similarly, because of the hierarchical aspects of skill it is impossible to lay down rules which will identify the level at which the analysis should be conducted or terminated. There are operational criteria which can be applied depending on the purpose of the analysis. For example, if it is concerned with the design of a training scheme, then one criterion will be that the analysis should be taken to a level such that skills to be acquired by selection are separable from those to be inculcated by training (Singleton, 1968a). The skills for which a training scheme is to be designed can then be looked at in further detail.

Skill description

The development of the specialized concept of skill took place during and after World War II stimulated by the experimental study of skilled operators in real situations (Singleton, 1968b). Two quotations from Bartlett (1943), the main originator of these ideas, should suffice to convey the essence of the concept: responses are not simple movements but 'coordinated actions the constituents of which can, and frequently do, change places'. Stimuli are – 'not a repetitive succession but a field, a pattern, an organized group of signals capable of changing their internal arrangement without loss of identity as an organized group'.

Skills are concerned with the organization of inputs and outputs. The exertion of force is not a skill in itself but the controlling of direction and position and graded application does require skill. Similarly, visual or auditory reception of stimuli is not skilled until these inputs are selected and organized into perceptual units. The difficulty about the term *skills analysis* is that it does not strictly cover the input and output mechanisms although they are essential to the practice of any skill and to the understanding of any job performance. Thus, such factors as vision and muscular strength are for operational purposes included in what is called a skills specification for a job. Considering the output organization, this is manifestly essential in all work because of the complexity of the skeletal and muscular systems which must be controlled and because of the associated problems of stability. It is easy to overlook the universal problem that an operator must not only do what he has to do, but he must do it without falling over or otherwise getting into unstable positions. These output skills are distinct from the input skills because of the specialized kinaesthetic system of proprioceptors and other sensing units which monitor the sequencing and success of outputs (Figure 18). It is convenient to postulate the existence of an internal effector organization system containing these output subroutines which can be reeled off on a command in the form of a decision and which can be controlled at least in the first instance by the kinaesthetic system. These are motor skills. When we talk about the development of a motor skill we mean the building up of an output routine of this kind. The lowest levels of motor skills are those controlling simple joint movements but the hierarchy builds up into the extraordinarily complex routines used in ball games and in playing musical instruments.

The input organization appears superficially to be analogous in that there is an integration of data through different sense modalities and there is a kinaesthetic aspect in, for example, the maintenance of stability of the visual world in spite of eye movements. Nevertheless there are fundamental differences in that the input organization system – usually called perceptual organization – is concerned with models rather than with routines. The simplest kinds of models are

a) Coordination of outputs

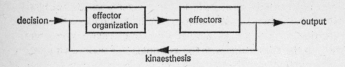

b) Coordination of inputs

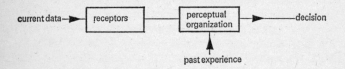

c) Four-block

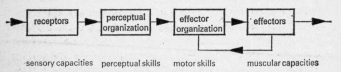

Figure 18 (a) Coordination of outputs (b) Coordination of inputs
(c) Block man

spatial which correspond topologically to the real world but
the higher levels of model are symbolic and abstract involving the techniques of languages and mathematics. When
we talk about the development of a perceptual skill we mean
the acquisition of a model into which sensory data is fed and
from which predictions which lead to decisions are possible.
The situation is summarized in Figure 18c. Skills analysis is a
problem of identifying required inputs and outputs and the
perceptual and effector skills which utilize these inputs and
outputs and which combine to provide the human operator
with the facility of the dexterous decision maker.

Skills analysis

1 Output aspects. Motor skills are best approached through
the experimental control of outputs. One method is to repro-

duce the situation in a simplified form. This is not easy to do and all methods must include some check that the experimental situation has validity in relation to the real situation, by demonstrating that skilled operators retain their skill within the laboratory task for example. It is then possible to manipulate the effectors so as to acquire evidence on the structure of the skill. For example, Singleton (1957) designed a laboratory task similar to a sewing task practised widely in shoe factories. He demonstrated that skilled sewing-machine operators performed much better on this task than did office workers. He then demonstrated that substituting a hand control for a foot control eliminated the skill, so also did substituting an off–on control for a graded control. These results showed that in the original hypothesis that the skill consisted essentially in an internal timing mechanism was inadequate; this skill at least depends on receptor and effector modes as well as on central processes.

Various other methods have been suggested such as examining timing so as to detect grouping of outputs, suddenly cutting-off inputs and measuring for how long outputs continue, etc., but they do not seem to have been used very widely.

2 Input aspects. This is only really feasible when the operator is using entirely artificial displays. In principle it is possible to cut off the sensory cues at the operator but in practice there is little that can be done. To cut off vision is usually too dangerous, audition is difficult to cut off completely and kinaesthesis is almost impossible to interrupt. If, however, the operator is relying on visual and auditory displays these can be manipulated and effects on performance detected. The use of computer-based displays which await operator demand are a recent development which has great potential for skills analysis. In these circumstances it is relatively easy to record what the operator asks for and when.

3 Achievement. This obvious measure is full of snags in practice. Overall achievement is only relevant to skill in a statistical sense, the best player does not always win even in a two person game. The method of achievement is very difficult to evaluate because in any skill worthy of detailed study, there are a large number of ways of achieving the desired end. For example, the complex skill of landing an aircraft was studied by measuring how closely the pilots kept to the optimum flight path. It was found that highly experienced pilots had a lower score than those with relatively little experience; presumably the former were good enough not to have to stick to the optimum path.

4 Tracing learning. Since highly skilled behaviour is such an integrated unanalysable activity one possibility which suggests itself is to examine the way it develops rather than the end product. This is subject to problems of individual differences, exposure differences, training differences and so on but there is potentiality here. Fitts (1964), for example, suggests that a skill might be analysed in terms of how this highly organized activity in space and time came about. He proposes that the investigator should look for *continuity* – spatially and temporally, *coherence* – that is relative redundancy, stereotypy and *complexity* – that is the size of the alphabet of stimulus-response transformations. Gagné (1965) has a similar approach but with a rather different taxonomy of *associations, multiple discriminations, behaviour chains, concepts and principles.*

5 Other taxonomies. Wallis (1966) suggests that skills should be subdivided into *procedural* and *routine* such as setting up and dismantling, control subdivided into *craft* skills which involve tools, *operating* skills which involve guiding and *coding* skills such as typing, *intellectual* skills such as reasoning and inferential thinking and *managerial* skills which involve leadership, decision making and committee activity. Fleishman (1966) has used factor analysis on performance data from a variety of laboratory tasks to arrive at the following dimensions: *control precision, multi-limb coordination, response orientation, reaction time, speed of arm movement, rate control, manual dexterity, finger dexterity, arm–hand coordination, wrist–finger speed* and *aiming*. It will be appreciated that his laboratory tasks were dominated by motor-skills.

Error Analysis

The human attitude to human errors is ambivalent. On the one hand we recognize that people are bound to make mistakes, on the other hand we regard individuals as free agents determining their own actions and therefore responsible for their own errors. This is not merely an abstract philosophical argument useful for an evening's conversation about free-will, its resolution is critical in determining procedures for all kinds of practical problems. For example: the appropriate punishment for guilty drivers in road accidents, how far a worker's compensation should be influenced by the cause of the error which resulted in his injury, what is a tolerable risk level in given circumstances and so on. Although it is possible to provide some evidence relevant to this kind of problem the evidence is never complete and value judgements inevitably enter in all decisions. This is why it is difficult to regard 'safety' and 'accidents' as legitimate objectives for scientific studies. Safety problems have legal, economic and ethical aspects as well as overwhelming and rather specific technical aspects. Superficially at least this leaves relatively little scope for the human factors specialist. Nevertheless, behind every safety problem and every accident there is the problem of people making mistakes. Thus, if there is a general supporting science for safety work it is the study of human error. Why people make mistakes is a question obviously requiring a multi-variate answer. We can make one attempt to reduce the complexity of the problem by distinguishing between different kinds of error. Thus, as in other aspects of human activity analysis the fundamental problem is to design an effective taxonomy.

Error description

There are many dichotomies for the classification of errors of varying relevance to system design. For example, there are so-called errors of commission which are positive in that the wrong thing was done, and there are errors of omission which are negative in that the right thing was not done. This is not a particularly useful distinction either conceptually or operationally. Logically it is not sound since these are not mutually exclusive categories. Furthermore two dimensions are confused; rightness–wrongness and action–inaction. Operationally the distinction does not assist either in allocating blame or in tracing causation.

Another distinction is between systematic errors and random errors. These are statistically distinguishable in that the former involve a bias which can usually be determined. Once the bias is known it can be corrected, at least in an average sense. Random errors can also be corrected by the 'principle of verification through independent duplication' (Chapanis, 1960). For example, if, in card punching, two operators independently punch the same data and their outputs are compared automatically, assuming that the comparator is reliable, the only errors which will get through the system are those where the same error has been made by the two operators. If necessary, the independent systems can be triplicated or even quadruplicated as in some aircraft systems, thus making error rates negligible providing the systems are entirely independent. This proviso is not as easy to achieve as it might appear. Human operators do have consistent error tendencies such as simple reversals, and hardware systems are often susceptible to common insult, for example an aircraft can be struck by lightning. There is an important operational distinction between detected errors and undetected errors. Providing errors are detectable it is usually not difficult to provide error-correction systems; if a tape punch operator has a back-space and erase facility, the system will be faster and freer of output errors. Within computer systems it is common practice to classify errors as programme recoverable, human recoverable and non-recoverable.

In an attempt to trace causation in human error Kidd (1962) has proposed a five-fold classification depending on

Table 8 **Error Taxonomies**

Origin	Taxonomy
Classical	Commission – omission Reversible – irreversible
Computer technology	Detected – undetected Programme recoverable Human recoverable Non-recoverable
Chapanis (1960)	Systematic – random
Kidd (1962)	Failure to detect signal Incorrect identification Incorrect weighting Incorrect action selection Commission
Meister and Rabideau (1965)	Failure of performance Incorrect performance Out of sequence performance Man required performance
DeGreene (1970)	Inputs: sensing, detecting, identifying, coding, classifying. Outputs: chaining, omissions, insertions, misordering. Decisions: estimating, logical manipulation, problem solving
Blame allocation	Incorrect diagnosis Insufficient caution or attention Wrong procedure Wrong instructions

the stage within human information processing where the blockage is likely to have occurred. That is, failure to detect, failure to identify, failure to weight the evidence accurately, failure to select the correct action and failure to act. Such an analytic system leads logically to improved interface design, task design and training design. Meister and Rabideau (1965) starting from the theory of failure of skilled performance, propose a four-fold classification. Performance may fail because no action is taken, the wrong action is taken at the right time, the right action at the wrong time or something unnecessary is done. DeGreene (1970) follows Kidd but uses a three-fold category, inputs, decisions, outputs with sub-categories of each. These taxonomies are based either on tracing what happened when an error was made or on what can be done about it. A third approach is to try and answer the question; who or what is to blame? It might be the training, the operator, the drill or the supervisor depending respectively on whether the error was due to incorrect diagnosis, insufficient care, wrong procedure or wrong instructions. These again are not exclusive categories.

Table 9 **Factors in error minimization**

Interface and workspace design

Selection and training

Rigid procedures

Human and hardware based monitoring

Contingency planning

Hardware maintenance

Working hours and conditions

As in all kinds of total activity analysis it emerges that the optimum classification depends on the purpose of the analysis. Given that the general purpose or goal is error minimization the particular purpose or strategy might be any one of those listed in Table 9 or any combination of them. Indeed the more critical errors become, the more necessary it is to adopt the total strategy indicated by taking all these measures.

Analytical techniques

1 Statistical techniques. The collection of statistics about accidents has been pursued for fifty years or more without too much success. For most of this period insufficient attention was given to the taxonomic problem just discussed and there has not been enough effort invested to provide a reasonable sample of data. Accidents are rare and thus it is difficult to get a reasonable size of sample, specific accidents cannot be anticipated (otherwise they could be avoided), and thus the investigator is reduced to relying on participants' recall. This can be distorted for all the standard reasons known to students of memory with the additional overwhelming factor that participants will inevitably present a picture which minimizes their personal negligence. In general the connection between accidents and errors is difficult to establish; an accident is often due to a combination of errors, but otherwise the statements which can be made apply only to particular kinds of systems or situations.

When large samples are used the multitude of relevant factors of unknown weighting is such that deductions can only be made on the basis of the theory of frequency distributions. Particular distributions are known to have particular predisposing circumstances. In their study of bus-drivers, Cresswell and Froggatt (1963) were able to demonstrate that their data most closely corresponded with a Neyman type A distribution rather than a negative binomial or a Poisson distribution. This suggests that all the drivers were at a high risk level for part of the time rather than that some of the drivers were at a high risk level for all of the time, or that risk is equally probable for all, all the time.

2 *Critical incidents.* One way of increasing sample sizes and reducing the distortions of reporting due to negligence avoidance is to study the accidents which might have happened but did not. The first extensive study of this kind is that due to Fitts and Jones (1947). They analysed pilot replies to questions of the kind: 'describe an error in operation of a cockpit control-reading or interpreting instructions which you have made or seen someone else make'. These reports were analysed into the categories shown in Table 10 and from these data it was possible to make recommendations about particular displays and controls and about the principles of

designing these interface elements — both separately and within consoles.

Table 10 **Error taxonomies used by Fitts and Jones (1947) for pilots**

Controls	substitution
	adjustment
	forgetting errors
	reversal errors
	unintentional activation
	unable to reach
Displays	misreading
	reversal errors
	interpretation
	legibility
	confusing instruments
	inoperative instruments
	scale interpretation
	illusions
	forgetting

3 Observational methods. In spite of the difficulties due to the rarity of accidents the method of on-the-spot study has been used on large systems. The investigator patrols or waits at some reporting centre such as a first-aid room and when an incident occurs he goes to the site immediately and tries to find out what happened by a combination of observation and interview. This has been used on road accidents (Hobbs, 1967) (in fact all police forces do this, but not necessarily in a scientific spirit) and on industrial accidents (Powell *et al.*, 1971). Another possibility is to conduct a task, job or skills analysis but with particular emphasis on potentially hazarduous situations. For example, Dunn (1971) has studied forestry work in an attempt to devise a technique of describing what workers do in a way which reveals stages or points where errors leading to accidents are most likely to occur.

4 The Maintenance Operator

Discussion in the previous chapters has assumed implicitly that the operator is on-line, carrying out his controlling activities in relation to the system objectives. This, in fact, is only one of the three kinds of function performed by the human operator: setting up the system, controlling it on-line and maintaining it. The setting-up function does not require extensive separate discussion. It is the easiest to analyse in that it usually involves a fixed sequence of actions. There are many tasks where the setting-up aspect requires greater skill than the on-line operation, e.g. the skill in cutting grass with a scythe lies in sharpening the scythe rather than sweeping it through the grass. High productivity in assembly jobs depends on the way the components are laid out rather than on high movement speeds in putting them together; in machine-tool work it is acknowledged that the setter is more highly skilled than the operator. There is a nice American phrase for these functions – the 'GOG aspects' of system operation, GOG being the abbreviation of 'getting off the ground'. It has already been mentioned that these cannot be automated in practice or even in principle since the GOG phase is intimately related to the system objectives which must be human orientated.

Turning to maintenance, there is another expressive American aphorism: 'any system which can function can also malfunction'. Maintenance is an inherent problem in all systems. The principles of 'allocation of function' apply but it is always more difficult to automate maintenance operations than on-line operations. The cost–value ratios are more discouraging since maintenance is relatively infrequent and it is therefore less easy to justify high capital expenditure. The variety of maintenance actions required is usually such that manipulative abilities are important and here the operator has a natural advantage over hardware. Detection of faults involves separating signals from noise which depends on pattern detection, whether in listening to the sound of a car engine or in looking at complex waveforms on an oscilloscope. Again the operator is much superior to the hardware. Fault-finding techniques often depend on putting together apparently isolated symptoms and arriving at a generalization about underlying causes. This is inductive reasoning for which hardware is not well suited. The operator's well developed error correction routines which check his own actions provide a sound basis for maintenance activities since they reduce the possibility of generating the unresolvable situation of a mismatch but no error identification, for instance if one consults two clocks which do not agree how does one decide what time it is? This can occur in automatic maintenance when there is a mismatch between the prime equipment and the maintenance equipment. It is not always obvious which one is malfunctioning. These, however, are relatively peripheral reasons for using human based maintenance. The main reason is that the designer is not infallible, not only is he incapable of designing permanently fault-free equipment, he is equally incapable of completely specifying the nature of faults which might occur. Thus he or another intelligent operator must be available to cope with unpredicted maintenance problems when they arise.

There is, of course, another side to the picture. Many maintenance routines are literally routine and the primary factor is that they should be done reliably; machines of course are better from this point of view. Human operators are expensive and unreliable. For most systems they cannot be held permanently accessible to be called as soon as maintenance is needed. Human maintenance operators also tend to be either over-optimistic or arrogant and to attempt tasks beyond either their capabilities or their equipment, thus making the prime equipment worse rather than better. The armed forces have faced an increasingly difficult situation ever since World War II in that the equipment has got more and more complex while the ability of the average maintenance engineer has decreased equally regularly. In the late 1950s these aspects stimulated a swing towards the hardware in allocating maintenance functions. This occurred for missiles, aircraft, radar and computers. After some years of experience the American Services decided that the situation

Table 11 **Human operator functions**

Setting-up or GOG (getting off the ground)	Getting equipment in the appropriate state for a given on-line function
On-line	Monitoring progress and taking appropriate actions in relation to specific objectives
Maintenance	Incorporates $\begin{cases} \text{servicing} \\ \text{fault finding} \\ \text{repair} \end{cases}$

warranted a large scale investigation which was conducted by the Rand Corporation (Swain and Wohl, 1961). This reached the following conclusions about automatic check-out equipment (ACE) generally.

1 The engineering arts had been pushed too far.
2 There was a gap between the practical and theoretical performance of the check-out systems.
3 The check-out systems were low on dependability in terms of both maintainability and accuracy.
4 The check-out systems were not sufficiently adaptable, e.g. changes in prime equipment usually required changes in check-out equipment but these were often not made. In addition the prime equipment could develop unpredicted faults not necessarily within the scope of the check-out equipment.

5 The check-out equipment was often not acceptable to the maintenance personnel who felt that their skills were not being sufficiently respected, e.g. some equipments were christened 'idiot boxes'.
6 Reliance on check-out equipment led to undesirable neglect of maintenance staff training and manual updating.

The conclusion was that 'the attempt to reduce man's participation in test and check-out equipment to the role of a monitor of a gadget is a fundamental mistake'.

In more general terms it seems that, as happened in other design areas, practice got rather ahead of principles and it was necessary to reappraise the problem. This led to an extension of the allocation of function methodology which, for these purposes, must incorporate procedures and check-out equipment as well as personnel and prime equipment (Figure 19). The whole methodology can now be reconsidered under these separate headings of system aspects, prime equipment, check-out equipment and procedures.

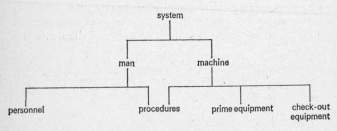

Figure 19 'Allocation of function' structure for maintenance operations

System Aspects

Design decisions in the context of maintenance are inevitably formulated on inadequate data. In every system faults will occur which are not foreseen at the design stage; even for those which are foreseen estimates of their frequencies will contain a large element of guesswork. These statements apply equally to malfunctions of hardware components and to errors of human components. It follows that every system design must incorporate flexibility to allow adaptation as experience develops and in particular there must be provision for feedback of information to designers, managers and operators about system reliability and causes of unreliability.

Prime equipment

The basic data required concern failure rates of components both separately and when functioning within subsystems. Unfortunately these data are not easily accessible. Just as there are large individual differences between human operators so the variation of component life, even of apparently identical components, is considerable. The operational research approach to predicting and avoiding malfunctions has had little success because of the complexity of the functions involved. For example, Figure 20 shows schematically a typical curve of probability of failure against operating time. There is an early period in which failure rate falls as components which were faulty when incorporated are eliminated, there is then a rapid rise in failure as components are fully stretched for the first time. As these are eliminated the

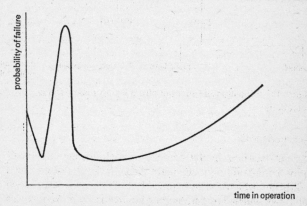

Figure 20 Failure in operation of components and sub-systems

failure rate drops again and then the true system decay curve becomes apparent.

One approach to design is to build for equal life, that is, to deliberately use components which will last about the same length of time. This is usually a legitimate objective although it is sometimes abused using the term 'built-in obsolescence'. There is no particular virtue in using a component that is of higher quality than that required since usually this will cost more. Again this applies to operators as well as to hardware. One example of equalizing useful life is in shoe components. These used to be such that an upper had the life of about three soles, but when entirely synthetic materials are used it is not difficult to arrange that the sole will last just about as long as the upper. When this principle cannot be applied because of obvious differences in useful life times then components with short lives should be easily replaceable and their rates of deterioration should be detectable, as for example, tyres on motor-cars.

Check-out equipment

In spite of difficulties in earlier attempts to design automatic check-out equipment there is obvious scope for their development. The criteria shown in Table 12 need to be balanced carefully, broadly by allocating as much routine as possible to hardware but leaving decision making with the human operator. One technique now coming into use for sophisticated equipment is to design prime equipment so that maintenance is not possible without special maintenance equipment, for example lifting gear and dismantling tools, and to deliberately restrict this maintenance equipment to those qualified to use it.

Procedures

These fall into three classes: servicing, fault-finding and repairing. Servicing problems are surprisingly uniform across all kinds of equipment. Mechanical components require lubrication and tolerance adjustment, electrical components require voltage adjustments. There is considerable scope

Table 12 Allocation of function criteria for maintenance operations

In favour of human operator
1 Low capital cost and versatility desirable since maintenance is infrequent
2 Wide range of manipulative abilities
3 Good at detecting signal patterns and signals in noise
4 Can apply inductive reasoning for fault location
5 Self-checking abilities
6 Can apply intelligence in unforeseen circumstances

Against human operator
1 Reliability needed in servicing routines
2 Continuous availability required since fault incidence is unpredictable
3 Hardware programmes can use designers' skills which are usually higher than field maintainers' skills

here for the design of training schemes, job aids and work-place layouts but the principles are exactly the same as for other man–machine interactions. Fault-finding generates interesting problems of strategy. A fault implies that a given input does not result in the expected output. The inputs must be present before the existence of a fault is postulated, for instance, running out of petrol does not imply a fault in a car. There are three ways of setting about tracing a fault. One can start either at the input or output end and follow a path through the system until the point is located where the appropriate signal disappears or appears respectively. A 'signal' might be the absence or presence of a voltage or of stuff or of information. One can check in the middle thus determining which half the fault is in, then check in the middle of the half and so on until one converges on the fault. One can examine the area where the fault is most likely to have occurred, then the next likely and so on until it is located. These are known respectively as the sequential method, the split-half method and the probability method. It is important to stress that terms such as 'path', 'area', and 'middle' are used above in a functional rather than a spatial sense. For example, if a motor car is assumed to be half electrical and half mechanical and the mechanic first checks whether he has an electrical or mechanical fault he is using a split-half method. Each method has its advantages. The sequential method requires the least mental effort and this can often be an important criterion because the fault-finder is frequently under stress; when a driver is trying to decide why his car has stopped on a road junction or when an operator of a weapon system is trying to do something about a fault in the presence of the enemy. Under these conditions well-practised routines are required. In fact deciding on optimum choice of strategy in different circumstances is one of the key tasks for the designer of training schemes for maintenance personnel. The probability method is not easy to train for in any formal way but it can be used very successfully by highly skilled and experienced operators. Usually training methods are based on the split-half technique with the assumption that following experience the maintainer will gradually switch to the probability method. The split-half is the logically superior method particularly for very large systems. The way in which its efficiency increases relative to the sequential method as the system size increases is shown in Table 13. In practice it is not usually possible to follow the 'split-halves' very precisely, but the method is not very sensitive to departures from exactness. That is, if one identifies a quarter where the fault is and three quarters where it is not this is a good start. However, success does depend on sticking to the rules and avoiding what may look like short-cuts. Taking short-cuts results in a mixture of probability and split-half techniques, which is not usually desirable unless the operator is very clear about when he is doing which.

Repairing

Two important problems for the system designer are deciding where particular repairs should be done and generating optimum instructions for repairers. For most systems there is a chance of doing repairs on site, at depot level, at

Table 13 **Comparison of sequential and split-half maintenance techniques**

No. of units	Average time to locate fault (minutes)	
	Sequential	Split-half
4	2	2
8	4	3
16	8	4
64	32	6
1000	$8\frac{1}{2}$ hours	10 minutes

Assume that the hardware is a single chain of units
and that it takes one minute to make one check

main base level or at factory level. This may sound like
military jargon but it applies equally to domestic equipment.
For example, the designer of a television set should decide
what can be done in the home, in retail shops, in regional
service stations and back at the factory. Such decisions are
necessary at the design stage since they affect size of units,
tools, accessibility, location of component stores and so
on.

The design of repair manuals or sections of manuals is still
relatively neglected. The designer often assumes, with disastrous results, that all these problems and even decisions
can be left to untrained technical writers.

The whole maintenance field is worthy of extensive design
effort. In most large technically based systems, for example
an armed service, the budget for maintenance including
spares and personnel will be greater than the budget for
prime equipment and on-line personnel.

5 Selection and Training

Training and education are procedures for the direction and encouragement of learning. Learning is evidenced by an increase in the ability to adapt effectively to the environment and more specifically, to perform particular tasks more effectively. There is extensive discussion in the literature about the distinction between training and education, much of it implying that they are qualitatively different and that training is somehow the inferior of the two. It is suggested that they must be kept separate or, at best, that there are just a few trivial aspects of education which can be dealt with by training methods. From a systems view-point the converse is true and education becomes a specialized kind of training. This point will be enlarged upon later but in this chapter anything said about training ought to apply equally to education.

Objectives

The first necessity in the design or appraisal of any training scheme is to clarify the objectives. It is invariably necessary to separate two levels of objectives: the general operational ones which are set by the parent system and the specific ones which are set by the needs of the system.

The general operational objectives are important because all training is parasitic in that it depends on resources provided by the parent system. It is not usually possible to demonstrate unequivocally that there will be a given return on this investment. There are three possible objectives:

1 to increase ability
2 to reduce learning time and–or cost
3 to decrease the required capacities.

They are usually thought of in this order but current trends in system design suggest that the reverse order now indicates relative importance. Training is commonly thought to be aimed at increasing ability or efficiency on the part of the individual, but if the work space and flow of information are properly designed and the operator is reasonably motivated he will eventually optimize his own performance by experience. The only thing that training can do in these circumstances is to reduce the time it takes to get to the highest level; this is the second objective. This presupposes that the task is not beyond his abilities; if it is then the third training objective becomes important. How important it becomes depends mainly on the quality of the available supply of operators in relation to the complexity of the task. In recent years there have been many cases where this is critical. Broadly, as automation increases the tasks required of the operator become more complex (this applies to the ordinary citizen as well as to the factory worker or serviceman), and it is therefore necessary to pay more attention to training to ensure that the operators can continue to function effectively. This aspect of training is inseparable from problems of interface design and job aids design discussed in later chapters. The specific objectives are set by:

1 the task description
2 the needs of the parent system.

The Task Description has already been discussed, its contribution to the design of training is fundamental, in fact this is one of the main reasons for taking the trouble to acquire it in the first place. The needs of the parent system are less easy to define but are nonetheless crucial. As Crawford (1962) puts it 'the training process is a creature of the system it serves'. A training designer who attempts to work without carefully considering the demands, standards and fashions of the parent system is asking for trouble. At its most obvious this is a matter of working within a budget which will be acceptable but there are equally powerful, if unstated disciplines, which include aspects such as using the kind of training methods which will be broadly approved, and working towards ends which conform to the parent system policy. For example, headmasters of schools usually try to avoid parent or parent-teacher committees because they are intuitively aware of the strength of these forces. Evaluations of industrial schemes which have failed often reveal that the failure was not because of any technical fault but because 'it didn't produce our kind of worker' or 'these people (products of the training scheme) didn't really fit in with the company image'. Another way of making this point is to suggest that training is always as much to do with attitudes as it is with skills. Training is essentially aimed at changing people and the parent system is properly concerned about the ways and directions in which they are being changed.

Selection

The efficiency of any training scheme is closely related to the homogeneity of the trainee population. This can be established by the Selection Specification which is extracted from the Job Specification. The aim of selection is to ensure that entrants to training have two characteristics in common:

1 The capacity to complete the training scheme and
   eventually to achieve 'experienced worker standard' at the job.
2 A set of basic abilities and skills which will form the basis
   on which to build task proficiency.

Selection is therefore essentially a problem of assessing capacities and abilities.

These can be subdivided into four categories:

1 Physical and sensory:
   Vision, hearing, size, muscular strength, handedness and so on.
2 Motor coordination:
   Locomotor dexterity, speed of response, timing and integration of hand and foot movements, etc.
3 Mental:
   Verbal and arithmetical abilities, manipulation of shapes, deductive skill, creative skill, etc.

4 Personality:
  Drive, initiative, reliability, tolerance of routines,
  resistance to stress, social skills, etc.

It will be noted that, in moving down this list the qualities described become more and more difficult to measure and also become more dependent on experience, training and upbringing rather than on native endowment. In fact the categories are partially determined by problems of measurement in that the physical and sensory capacities can be measured accurately by uncomplicated physical techniques, (size and weight) or at worst by psychophysical measurement, (visual acuity). Motor coordination assessment is essentially performance measurement (speed of manipulation and problems of motivation begin to complicate the picture). Mental measurement is the domain of what is called psychological testing (using paper and pencil methods, intelligence and verbal reasoning tests). The testers have extended their interests, not yet very successfully, into the personality sphere. This effectively remains a problem for the interview.

The debate on tests versus interviews as a selection technique is largely spurious for two reasons; one conceptual, one practical. Conceptually the two are not distinct and the difference is not expressible along one dimension. At the extremes, the test is highly structured and written, the interview has a low formal content and is verbal. However, tests can be verbal instead of written, the interview can be highly structured instead of open ended. In practical terms there is nothing to be gained by disparaging the interview, there is no alternative as a selection vehicle. Obviously the interview situation is susceptible to improvement. Study with this objective in view is potentially worthwhile but it is not susceptible to abolition. It is true that in some situations with a very high selection rate (ratio of number of candidates to number of places), for example in the entry of students to some university departments, the interview has been abolished. But this is merely a confession of selection failure.

Because of the limitations of assessment methods there is some justification for structuring the problem in terms of operational utility rather than fundamental conceptual validity. For example, (Singleton, 1970) the target search categories can be divided differently from above in the following:

1 Capacities:
   Physical and sensory characteristics, motor coordination at the elementary level and mental capacities.
2 Skills:
   Motor skills which are complex and specific output routines, perceptual skills which are specialized models for combining and interpreting input data.
3 Disposition:
   Personality variables such as punctuality and reliability.
4 Knowledge:
   Unidentifiable aggregates arising from education and experience, e.g. what a scientist knows or what an accountant knows.

These are not fundamentally separable, e.g. knowledge is made up of skills and dispositions, but are useful search terms and descriptive terms.

The approach based on facilitating the collection of evidence is best illustrated by the extensively used seven-point plan (Roger, 1952). The headings used are:

1 Physique – health, vision, hearing, etc.
2 Attainment – educational and occupational achievements
3 Intelligence – general ability
4 Special aptitudes – e.g. music, drawing, arithmetic
5 Interests – social, intellectual, recreational, etc.
6 Disposition – character traits, temperament
7 Circumstances – living, working, transport conditions.

The difficulty of applying any kind of universal metric to the description of an individual together with problems of knowing what is required anyway (the job specification) is such that selection is very much a probabilistic process. It follows that it is desirable to delay the final decision for as long as possible – the early stages of training are often the best sources of selection information – and to build in validation procedures as an inherent part of any selection method. The mechanics of selection are of great importance but they are based more on common sense than on science. Consideration of the physical environment of interviews and tests, accurate scheduling, requests for references, expertise in communicating with the potentially suitable population by advertising and so on can make or mar any selection scheme. The design of job specification data and application forms are discussed later.

The design of the training scheme

The discrepancy between the Job Specification and the Selection Specification indicates what has to be done by the training scheme (Table 14). At this stage it is necessary to consider how the operator will do it, that is what kind of model the operator will use in considering the problems set by the task. There are two extremes expressed by the terms programmed operator and concept operator.

Table 14 **Training Design Procedure**

1 Obtain training requirement from discrepancy between selection specification and job specification

2 Consider kind of operator required; whether programmed or concept or intermediate

3 Determine on-line/off-line separation

4 Consider methods of providing practice

5 Devise methods for progressive combination of skills

6 Consider methods of implementing and evaluating the scheme

Programmed versus concept

A programmed operator is one who simply has to follow detailed instructions, a concept operator is one who understands the principles behind the task and works out his own detailed instructions. For example, a boiler attendant might be trained to keep a pointer on a red line by turning a control: he merely needs to know that anti-clockwise turns will reduce the reading and clockwise turns will increase it. In these conditions he is a programmed operator. If, however, he is told that the pointer indicates pressure in a boiler which has an optimum value but this value can be adjusted by increasing or decreasing the rate of fuel supply, then he is a concept operator. Clearly the kind of operator, or rather the kind of model to be used by the operator, is a matter requiring an early decision and this decision will have a profound influence on the design of the training scheme.

Programmed training, sometimes called Stimulus-Response (S-R) training requires a Job Description in which the complete range of possible stimulus patterns is enumerated and the appropriate response set to each pattern is determined. The training consists of reinforcing the S-R links. Training differs from on-line performance in that additional knowledge of results is provided to reinforce correct responses and the stimulus patterns may be simulated. It is thus possible to make them more readily identifiable and to provide practice which is proportional to relative importance rather than relative incidence on-line. In an ideal world such training would be entirely unnecessary since all tasks in this category would be done by machines but there are often factors, for instance complexity of stimuli or capital cost, which result in the allocation of such tasks to human operators.

Concept training is dependent on the principles of human skill. The changes with increasing skill can be tabulated as shown in Table 15. The purpose of training is to facilitate the transition from the unskilled to the skilled mode of behaviour. For all processes this requires exposure to the kind of activities relevant to the real job with more precise knowledge of results and with frequencies adjusted to include criticality. That is, extent of practice in training should depend on how critical or otherwise important a task is as well as how frequently it is met within the real job.

On-line versus off-line

On-line operations are those involving real inputs and outputs of either material, power or information. For example, if a machine tool operator is using the ordinary materials of which products are made and if the products he makes are to be used then he is on-line, if an air traffic controller is dealing with data from real aircraft and if his decisions affect their subsequent behaviour then he is on-line. If either inputs or outputs are unreal then, of course, the operator is off-line.

There are considerable advantages in off-line training:

1 It may be less expensive, e.g. if components are made from materials cheaper than the real materials.

2 It may be less dangerous, e.g. if outputs are unreal then other people cannot be affected.

3 The objectives are more obvious. An all pervading snag in on-line training is that there is a conflict between training requirements and operational requirements.

4 The relative exposure of the trainee can be weighted by complexity and criticality of tasks rather than by their natural frequency of occurrence.

Provision of practice

The next broad decision is to do with methods of providing practice for the trainee. The basic problem is whether whole or part methods should be used. The continuous and interactive nature of human skill would suggest that whole methods are preferable since part methods inevitably create problems of how to link the parts and, more basically, whether the part outside the context of the whole is the same task anyway. On the other hand part methods have the enormous advantage that the possibility exists to provide more specific knowledge of results. Research on feedback in relation to training has been comprehensively summarized by Annett (1969).

Off-line part methods depend on training devices. These

can be divided into categories in various ways. Biel (1962) for example separates them into: automated teaching devices, concept trainers, skill trainers, procedure trainers and simulators. Although such training devices can be valuable there is a tendency to over-emphasize their importance partly because they provide a useful form of displacement activity for trainers – it is easier to concentrate on designing

Table 15 **Difference between skilled and unskilled operators**

Unskilled	Skilled
Output	
Unselected choice of method and medium	Sensitive choice
Order confused, timing variable, accuracy insufficient or over-sufficient	Order, timing and accuracy appropriate
Input	
Important cues have not been identified or structured	Relevance is established, cues are known
Much irrelevant data are considered	Other data are ignored
Excessive use of visual system	Less reliance on vision
Control	
Conscious attention	Less conscious attention
Apparent effort	Little or no effort
Frequent monitoring	Less frequent checks

a new device than to think carefully about trainer-trainee interaction. Some problems have arisen because the theory of these devices, that is of 'transfer of training' is inadequate. Because the optimum conditions of transfer are not easily deducible it is always tempting to make the training device as much as possible like the operational situation. This does have some justification in motivational terms particularly for older trainees who will see no sense in playing with a device unless it closely resembles the real hardware. The problem has been summarized by Miller (1962) as shown in

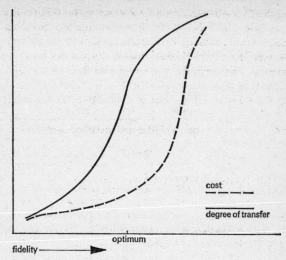

cost

degree of transfer

optimum

fidelity ───────➤

Figure 21 Fidelity in training devices

Figure 21. This is a schematic diagram, in practice neither fidelity nor transfer is easily measurable and yet knowledge of this kind is required before training devices can be justified. This is, of course, an allocation of function problem between trainers and devices and, in common with all such allocation problems there are no easy solutions and there are tendencies to rely too much on hardware.

The main kinds of devices used for training of machine operators are shown functionally in Figure 22. For perceptual skill training the patterns of likely stimuli with different noise levels are made available in standard programmes, for example for radar operators. The trainee must indicate that he has correctly detected the relevant signals within an acceptable time. His performance is checked by the trainer. For motor skills the trainee has to practise the correct pattern of movement at the correct speed, for example a setting-up routine. Again standards are usually available directly to the trainee but also to the trainer. For perceptual

perceptual skills

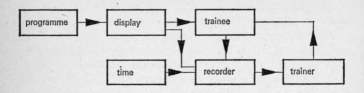

motor skills

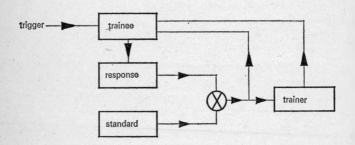

perceptual motor skills

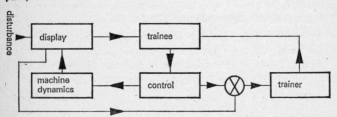

Figure 22 Individual training

motor skills (these are often called just 'skill trainers') a simulation of the appropriate machine dynamics is used within the loop and the trainee must respond to errors disturbing the system, for example all tracking tasks are in this category. His success is recorded by comparing display and control information and this is available to the trainer. In all cases the adaptive nature of the training system may depend on the trainer or there may be a feedback loop which does not include him and the system becomes self-adaptive. The important point is that, as a learning system there is no difference in principle between the two methods. In other words, from the point of view of principles of learning and training there is nothing unique or even novel about self-adaptive training systems.

Progressive combination of skills

The 'progressive part' method of training (Seymour, 1968) utilizes the advantages of unambiguous knowledge of results in that the sub-tasks are first learned separately. The method avoids or at least reduces the problems of integrating the separate parts by doing this progressively (Figure 23). For example (Singleton, 1959) a training scheme for shoe machinists involved the following states (Figure 24).

1 The basic skills of positioning, guiding, machine control and machine servicing were practised separately using a variety of training devices for each task.

2 The positioning, guiding and machine control skills were then combined into stitching exercises without thread.

3 The machine servicing skill was then added by using stitching exercises with thread beginning on training curves and progressing to simple shoe operations on synthetic material cut to resemble orthodox shoe components.

4 Technological knowledge was added by lectures and visual aids and this was combined with the manual skills by carrying out the last stage of training using real components.

Implementation and evaluation

In all technological activities, developments absorbs at least ten times the resources required for research. This applies to

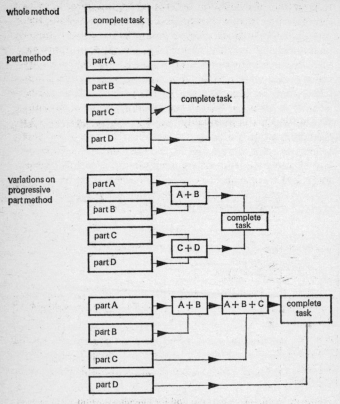

whole method

part method

variations on
progressive
part method

Figure 23 Stages of practice in training

training in that, even for an elaborate job for which highly
skilled operators are needed, the analysis of the skill and the
design of the training scheme can be done in a matter of
months. There must then be a period of years in which the
scheme improves and spreads slowly until the whole of the
relevant population are exposed to it. The effort required for
this phase is largely administrative and social but some
scientific effort is required in the sensitive design and in-

terpretation of evaluation procedures. It was mentioned at the beginning of this chapter that training is not easily susceptible to techniques such as cost–benefit analysis. For a manufacturing industry the benefits are often in improved morale which leads to operator reliability and better quality rather than simple increases in productivity. In service industries, improved training does not have effects which can be isolated from other factors contributing to customer satisfaction which will eventually reflect in the balance sheet of the company. It is often possible to demonstrate that a new scheme is better than an earlier one but there can never be any absolute measure of training success. For this reason there is always scope for improvement and the design of training becomes a continuous dynamic activity rather than something which can be dealt with just once.

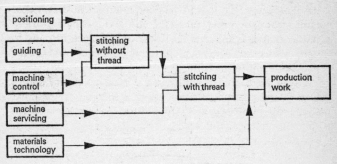

Figure 24 Progressive part method for machine sewing

6 Job Aids and Training Aids

Job aids are static information presentations. They differ from other displays in that they do not change in time and differ from training in that there is no interaction with the recipient. Nevertheless a job aid such as an instruction booklet for a washing machine is a form of training device and a job aid such as a lubrication chart for a maintenance engineer is a form of display. Thus job aid design is a problem intermediate between interface design and training design (Figure 25). It is part of the interface in that it contains information relevant to the task and it is part of training in that it is the simplest form of guidance on how to undertake a task. The job aid is to the human operator as the programme is to the computer (Wulff and Berry, 1962).

Given this degree of importance it is extraordinary that so little attention has been given, either in theory or in practice, to the design of job aids. In the history of computer design there was a quite rapid realization of the importance of soft-

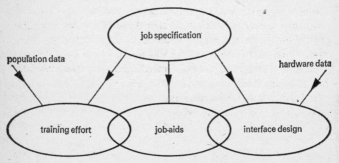

Figure 25 The central role of job aids in job design

ware as distinct from hardware; the former has now become the centre of attention and the more expensive in terms of design and development effort. Yet in the design of most man–machine systems there has been no corresponding movement to focus attention on job aids as an integral part of the system design problem.

Maker–user communication

One way of looking at the job aid problem is that it forms the communication channel between the producers of systems and the users (Figure 26). It is never enough merely

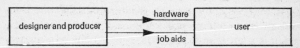

Figure 26 Job aids as a communication vehicle

to send hardware from the maker to the user. The user needs information about limitations, optimum working procedures, how to set up the system, and what to do when it malfunctions. In American terminology this information is provided by what is called the 'OJT package' (on the job training). This package may consist of no more than a sheet of paper (how to wire-up an electric toaster and what the controls do), it may require a whole series of manuals – as for a computer – or it may be a team of men and equipment employed by the maker and temporarily loaned to the user, e.g. on aircraft. Whatever its size there is always a design problem which requires considerable expenditure of time and money. Unfortunately, as Vandenberg (1967) points out, this kind of problem only becomes urgent at a relatively late stage of the design process by which time, almost inevitably, the budget is overstretched and the system is behind schedule. The basic difficulty, which is common to other aspects of systems design, is that the overall standard at national and international levels is so low. Users are not accustomed to good job aids, they do not expect them and they do not insist on getting them. Thus the management of production organizations have insufficient incentive, their competitors are as bad as they are, so there appears to be no problem. The remedy is for the customer to insist on a higher standard and for there to be an adequate conceptual background available so that the appropriate standard can be provided when it is demanded.

The design of manuals

The manual is part of a communication system; in terms of communication theory there must be a source, an encoder, a channel, a decoder and a destination. For the designer of the manual there are corresponding problems of data collection, classification, selection, processing and transmission. The destination is the user; his abilities to decode the information in the manual are the context for the encoding design problem.

One of the advantages of the systems approach is that the requirements for any stage are available from previous stages. If the design has not been carried out systematically then acquiring comprehensive data for the manual will be prohibitively tedious and expensive. If, on the other hand, there is already available a clear statement of the system specification, the hardware specification and the task specification, then there is no problem in generating data for the manual. There is, however, a specialized problem of encoding these data for this purpose.

The first problem is *classification*. This is usually best done in terms of tasks required of which the main categories are setting up, on-line operation and maintenance. A manual which has even this minimal structure of separation into three parts would be a considerable improvement on most of those currently available. There has to be a considerable data *selection* process to reduce the total information to a minimum; the selection criteria are available from the system objectives. Broadly the user needs enough information to ensure that these objectives are achieved without overloading the system. The task description provides a useful check that all necessary data for the decisions and choices are available.

Table 16 Manual structure – the graded introduction to the task

1 Definition of terms

2 Requirements and performance criteria

3 Orientation
(flow diagrams, charts, theory)

4 Procedures
(stimulus variable, response alternatives, action sequences)

5 Parts lists

The *processing* involves providing information required for the making of decisions in the sequence in which the user is most likely to make them. In order to *decode* all this data, the user needs a context and this can be provided by a graded introduction to the task with the structure shown in Table 16. As a start all the technical terms used must be defined; it is common when working in a design or production organization to develop an esoteric language which everyone within the organization understands and finds useful. It is easy to forget that outside the organization the terms may not be so well understood and may even have different meanings. Having established the language, the next step is to describe what the system needs and what it is capable of, in other words the system inputs and outputs. The stage is then set to examine how the system functions. This is best conveyed first in an overall and general way, using charts and diagrams, followed by greater detail on separate functions or components. Although some theory may be necessary, it should be kept to a minimum. Writers are too often carried away into writing a miniature text-book on the science and technology behind the system. The reader may find this of some interest but it is almost certainly not relevant to his function as a user. Similarly the designer may be particularly proud of the way he has solved some technical problems within the design and the manual will tend to dwell at length on these aspects. Again firm discipline is needed: the elegance of the design solution is not relevant to the needs of the user. *Procedures* are best described schematically and diagrammatically rather than in wordy texts. *Parts lists* always have problems of both identification and search. Words and letter codes can often be used to leaven the obscurity of parts numbers. (Currently it is quite common for these to have a sequence of eight to ten numbers intermixed with a few dashes.) Small schematic drawings of parts are an enormous help in searching through lists or tables.

Finally in considering the *channel* of communication the orthodox printed book can be usefully supplemented by legend plates, wall charts, films, film strips and even teaching machine programmes. If a new machine or range of machines are to be introduced on a large scale, then films and film strips can be a valuable aid to senior personnel such as the chief engineer or training officer in helping to convey information about the new systems to the personnel who are going to be involved with them. Similarly teaching machine programmes can be used to communicate with individuals or small groups who are otherwise difficult to contact. For example, they have been used to convey information to aircraft maintenance crews who are scattered throughout the world.

Static displays

Legend-plates and wall-charts are particular examples of this method of information presentation. Their most common purpose is to provide a machine operator with action sequences or with data which are used for machine settings. Easterby (1967b) proposes three possible approaches to the theory of this kind of problem. *Communication theory* and particularly the concepts of noise and redundancy. The principles of *language construction* with the distinctions between pragmatic, semantic and syntactic rules. The principles of *perception* including figure-ground relationships, stability, continuity and unity.

The application of communication theory to systems with human operators as destinations has been developed in detail by Garner (1962). In particular his distinction between structural and signification aspects of meaning has implications for display design. Structure which can be quantified in terms of redundancy is related to the Gestalt concept of figure-ground relationships. Signification is what the linguists call semantics. Linguistic distinctions form a valuable conceptual basis for the design of symbol systems. Clearly whatever set of symbols is designed must have some pragmatic rules; that is there must be the implication that they belong in a particular context. Semantic rules are to do with the relationship between the signs and external referents such as actions or objects. The syntactic rules govern the relationship between the signs themselves: a set of symbols needs its own grammar (Easterby, 1967a). Apart from symbols, static displays often contain many numbers. Again the structure of the tables of numbers, its relationship with the questions the human operator is trying to answer, and its indications of his optimal actions need to be considered systematically (Easterby, 1966).

Jobs aids for designer

The designer himself is a human operator and it is worth attempting to turn these design techniques back on themselves as it were and look upon the designer as an operator within a system with his own informational needs. The engineer of course has his manuals of data about strengths of materials, screw threads standards, conversion units and so on. There are analogous job aids on human operator aspects in the form of manuals of human factors data. They vary from small pocket-books (White *et al.*, 1960; Kellerman *et al.*, 1963) to large encyclopaedic volumes (Morgan *et al.*, 1963; Webb, 1962). Although they undoubtedly can be valuable in giving the designer some indications about human factors questions, they have serious limitations. In cases where specific numerical data are required, for example in body size measurements, even a large manual cannot contain all the detailed measurements which are required for all the possible populations which the designer may be considering. Non-numerical recommendations, for example that a particular display or control is 'better' than another, raise more questions than answers. For example, in what circumstances is it 'better'? What is meant by 'better'? How much 'better'? How much does it matter whether or not the 'better' solution is adopted?

This kind of criticism has led to the adoption of an alternative approach using check-lists. Here the principle is to provide a systematic series of questions which the designer can ask himself in the hope that he will be reminded of any relevant human factors which he may not otherwise have considered. One of the most comprehensive of these is that constructed by the International Ergonomics Association. It is reproduced in Edholm (1967). Within each section there are detailed questions such as 'are the characteristics of the hand controls compatible with the forces required to operate them (shape, size, surface) and are the forces acceptable?', and 'is the pointer simple and clear, and does it allow the numbers to be read without obstruction?'. Such checklists have been criticized on the grounds that they are essentially negative and error correcting; it is argued that it would be better to use systematic methods so that design errors do not occur in the first place. This is perhaps a counsel of perfection.

There are various aids to the designer which attempt to compensate for his limited ability to visualize and manipulate hypothetical situations. The classical approach of scale drawings has an inherent limitation in being essentially two-dimensional. Spencer (1965) has shown that even experienced draughtsmen and engineers have difficulties in translating information from a scale drawing into a three-dimensional form. For example, everyone finds perspective and isometric drawings easier to comprehend than either first or third angle projections. Such projections do have the virtue of greater precision and there are ergonomic aids such as scaled manikins with rotating joints at the appropriate points to represent human operators. These can be fitted on to scale drawings to check the adequacy of work space dimensions. More dynamic models of this kind which can be generated by computer systems have been developed. These reveal the limitations of available data of body size and available models of human movement. It is important to use manikins representing the large and small operators rather than the average ones. Attempts to design full-scale models of human operators have not been very successful; it is almost impossible for example to design one which has the flexibility of the human frame and yet does not slump into absurd postures. Thus these tend to be very expensive and their use is confined to dangerous testing situations such as ejection seats for aircraft. Nevertheless full size models of potential work spaces are extremely valuable as testing situations for human subjects. Chapanis has pointed out that the only satisfactory model of a human operator is another human operator.

Looking at the problem more generally the whole of this book can be regarded as a particular kind of job aid for a designer in that the main purpose is to provide a structure for design thinking. This can be regarded as the third strategy, the first being to provide data, the second to provide checks. This is not entirely accurate since Systems Design techniques deal only with problem identification. Further techniques are needed to solve problems, some dependent on factual information such as Human Factors data, and others dependent on the creativity of the investigator. Every design problem involving people is too unique to be dealt with by a standard procedure.

Training aids

Training aids, job aids and interface elements are all methods of presenting information. The term 'display' usually connotes an interface element conveying information on the current status of some dynamic system. 'Job aid' implies that static data relevant to a task are being presented. A training aid is like a job aid in that the information presented is essentially off-line, but although it can be static it need not be. The data presented are relevant to the building up of expertise about some task or set of tasks and the dynamic mode is often useful in this context. The range of training aids is shown in Table 17.

Table 17 **Training aids** (after Biel 1962)

Static	Dynamic
Tables	Films
Charts	Closed circuit TV
Nomograms	Teaching machines
Texts	Concept trainers
Filmstrips	Simulators

Tables of data require careful structuring for clarity of interpretation. The matrix should not be entirely regular, different divisions are required at intervals of not more than five items in either rows or columns. The sequencing of data should follow the lines of the operator's likely sequence of questions rather than the logic of the mechanism. If there are more than two independent variables, additional coding, often by using different symbols, is required.

In designing charts and maps the most common error is to stick too closely to uniform scaling. Human operators have no difficulty in interpreting different magnifications providing the structure corresponds to the structure of the external referent. For example, in indicating voltage test points, if there is a large cluster in a small area within the chassis, this area can be magnified on the chart without causing confusion providing the broad structure of the chart corresponds to that of the chassis. Nomograms and other aids to memory and computation need to be scaled according to the accuracy required and to contain no more data than is essential. The standard problem here is too much visual noise.

In the writing of texts there are corresponding structural

problems of division into sections, graphical versus verbal material, use of indexes and so on. Again the key to the design problem is to start from the needs of the operator as indicated by the Job Specification rather than the logic of the system. Film strips have their main value in dealing with groups of operations with a human trainer as the basic presenter of information. Films and closed-circuit television systems are particularly valuable in providing detailed but unobtrusive observation. The obvious virtue of the film is its potentiality for compression and magnification of events in time as well as in space. The closed circuit television is particularly valuable in conveying techniques of people interaction such as teaching and interviewing. After a decade of high fashion and exaggerated claims the teaching machine is now finding its proper place within the repertoire of teaching aids. It has two main advantages: self-paced learning and the fuller utilization of the really skilled teacher (in this case the programme-writer). There is no difference in principle from other teaching methods. The concept trainer is usually a hardware device which can be manipulated by the trainer so as to provide the trainee with experience of situations which may arise in practice and which can encourage the development of optimal working procedures. For example, in teaching fault finding in electronic circuits, standard circuits can be made into which various faults can be deliberately introduced together with cues on how to locate them. Skill trainers present the trainee with a task requiring correct choice and grading of responses to given stimulus situations and also provide detailed knowledge of results. For example, some so-called driving simulators come in this category. The trainee is required to operate controls and this causes things to happen on visual and auditory displays, but these displays can be totally unlike those met with in the real situation. Simulators combine the characteristics of concept trainers and skill trainers. In addition they are usually designed to provide conditions closely resembling those met in the practical situation.

Job aids versus training aids

There are often system design decisions which require a choice of either job aids or training aids. The task is such that it is possible to rely on a relatively unskilled, unknowledgeable operator, providing he is given the more comprehensive collection of job aids. The job aids support his skill and provide him with increased potentiality to deal with situations not covered by training. This technique of basic training plus job aids is satisfactory and economic in systems where there is no time stress and where errors are reversible. In this context the job aid becomes a kind of on-line training aid in that the operator will learn by the use of these aids, for example training for domestic equipment such as automatic washing-machines is achieved by this method. On the other hand, in situations where fast reactions are required and errors are dangerous or expensive then there is no substitute for extensive training including refresher and updating training by the use of training aids such as simulators.

7 Interface Design

An interface is an imaginary plane across which information and power are exchanged. In the case of the man-machine interface, the exchange is predominantly of information. Information passes from the machine to the man through the display elements of the interface and from the man to the machine through the control elements of interfaces (Figure 27). From the design point of view it is important to maintain

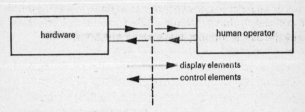

Figure 27 The man-machine interface

the attitude that displays and controls are not parts of the machine, they are links between the machine and the operator. There is a specific and difficult design problem because of the fundamental differences in performance characteristics of men and machines. This is the reason for the success of the man–machine combination, but it is also the source of serious problems in establishing efficient communication between the two parts. Starting from the human operator the problem can be divided into four categories corresponding to the four main functions within the human information processing system. The reception of information requires consideration in terms of the psychophysics of displays. The aim is to achieve the best match of the charac-

teristics of the information presentation with the characteristics of the human sense organs. The operator must
then organize the information inputs perceptually; the
process can be facilitated by appropriate structuring of the
presented information – this is the display encoding aspect.
When the operator has made a decision he must then organize the outputs to implement that decision. In particular he
must aim for the optimum sequencing and timing, in this
way he interacts with the machine dynamics. Finally these
decisions are effected by connection between the operator
and the machine through controls. Control matching for this
process is essentially a problem of functional anatomy.

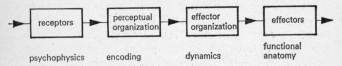

Figure 28 Human information processing and corresponding
interface design areas

The order in which these four categories are dealt with
depends on the particular interface design problem. It is
often described in the order of information flow as shown in
Figure 28. This gives priority to the human input problems
and there are usually the more important when dealing with
military systems. However, there is a case for reversing the
order of consideration on the grounds that it is the machine
and not the man which is the subject of design. The outputs
of the operator are the inputs of the machine and the Task
Description, which is the starting point of the interface
design, also begins with human outputs. In general the order
depends on the investigator's estimate of where the information blockages or bottlenecks are likely to be. That is on
whether input or output aspects are going to be most critical.
If there is no reason for assigning priorities from this point of
view, then it is best to start with anatomical problems, then
deal with dynamics and coding problems and finally with
psychophysics problems.

Functional anatomy

In using the limbs there is first a choice between foot and hand. The hand is superior for speed, accuracy and selectivity and the foot for application of force. Greater forces through the foot are more efficient partly because it is easier to utilize the body weight. In general, for the application of force, the dexterity provided by the many-jointed skeletal system is used to adjust the body to an optimum posture. The optimum posture is such that force is applied by utilizing body weight rather than muscular strength. Unfortunately applying the body weight through a foot usually demands a distorted asymmetrical body position with consequent problems of stability. For these reasons it is usual for a foot pedal operator to be in a seated position. This then requires careful design of the seat partly to take the reaction from the foot force on the pedals and partly because the feet cannot in these circumstance carry out their normal role of providing stability and taking a proportion of the total body weight. For example, in operating the heavy rudder bar of an aircraft, the design of the back rest of the seat is critical. In operating the large circular disc which spins the work on a linking machine in the hosiery industry, the design of the seat-pad is critical.

It is commonly assumed that it is better to work from a sitting position rather than a standing position, but this is not necessarily so. There are a number of general factors which favour standing work: the standing reach area is more than twice as large as the sitting reach area even assuming that the feet do not move. This provides a much greater space into which controls can be fitted. For any reasonable degree of comfort in sitting the legs must have a lot of space in front of the resting position of the hands (at least fifty centimeters). It is impossible to function at a vertical interface from a sitting position. The sitting posture precludes the use of body mass and momentum for the exertion of forces. The legs are very efficient damping devices, any vibration situation is easier to adapt to from a standing rather than a sitting position. When the operator must move away from the work-space at frequent intervals the effort of standing up and sitting down will require much greater energy expenditure than standing for the whole time. This effect can be reduced by using high seats and footrests. Incidentally, postures other than standing or sitting, kneeling or lying for instance, should not be used except for very short intervals For example, there were attempts to use pilots in the face-down posture so as to achieve an aircraft with a very low cross-sectional fuselage area but these proved abortive. There are impossible problems of head control as well as of limb actions.

On the design problems of foot controls there are four aspects always worthy of consideration. All these can be exemplified by the design of foot pedals in cars. Firstly the separation of controls should be adequate. This is partly to ensure that they can be operated separately – it is still possible to find cars where man wearing heavy boots cannot operate the clutch without also touching the brake – and partly to provide fast selection. Controls which are close together only require short foot movement times but they may well require much longer choice times because of the excessive precision of the selectivity. The foot is quite a heavy object and all return springs on foot controls should be adequate to support the weight of the foot (about 3 lbs). It is common in cars to provide very light accelerator controls on the grounds that it makes the car feel 'lively': the slightest touch and the engine revs. Unfortunately this means that in the cruising mode the driver has to develop muscular forces to keep the weight of his foot off the control and he will complete a long journey with a stiff leg. Thirdly, the position of the pivot is worthy of careful consideration. An effective point of rotation beneath the instep is usually a point beyond the toes, thus causing the foot to slide along the control surface as the control is operated; this can encourage the foot to slip, especially when the shoes are wet. Finally, returning to the question of selectivity, foot controls must be such that they can be operated without visual guidance. The dip-switch

All controls need clear identification and separation from others		
	use	special design requirements
button	 in arrays for rapid selection between alternatives	to avoid slipping finger and accidental activation
toggle	 for definite, rarely used action involving only choice of two (normally on/off)	to avoid excessive finger pressure or nail damage
selector	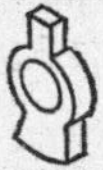 for more than two and less than ten choices	to avoid excessive wrist action make total movement less than 180°. do not use simple circular shape
knob	 for continuous variables	size depends mainly on resistance to motion use circular shape with serrated edge
crank	 when rotation through more than 360° is needed	grip handle should turn freely on shaft
lever	 for higher forces or very definite activity	identification of neutral or zero
wheel	 for precise activity involving large angles or rotation	identification of particular positions avoid slipping

Figure 29 Types of hand control

for headlights is often located almost at random to the left of other foot controls; this can be disastrous since it is usually required quickly and yet the foot must search for the correct position.

On the design problem of hand controls the major difficulty is in selecting the most appropriate kind of control. There are a large variety of controls commensurate with the enormous versatility of the hands. However, this means that it is difficult to choose between buttons, toggles, selectors, knobs, cranks, levers and handwheels. Some general criteria as indicated in Figure 29, details of optimum sizes, pressures, separation and so on are given in the data handbooks.

Dynamics

Within any man-machine system there are four components, each with its own dynamic characteristics (Figure 30). These can be described in turn:

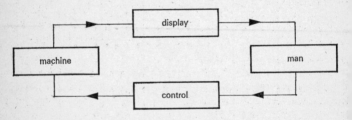

Figure 30 Man-machine components with different dynamic characteristics

Man dynamics. The main properties of man as a dynamics system are:

1 He is subject to a delay of at least half a second between inputs and related outputs.

2 He has very low precision as an analogue computing device. That is, he is very poor at addition, subtraction, differentiation and integration when compared with mechanical or electrical ways of carrying out these operations. For example, a cheap mechanical speedometer is between ten

and a hundred times more accurate than a human operator in assessing speed.

3 He is a source of noise mainly of two kinds: muscular tremor which is always added to his intended outputs and inattention which results in unexpected discrepancies between given inputs and expected outputs.

4 He has a highly adaptive gain control system and therefore a highly variable gain. That is, the ratio of output to input is entirely unpredictable unless the context is well understood.

For these and other reasons his transfer function cannot be complete. For the human operator there is no equation of the general form

$$\theta_0(t) = f\,\theta i(t)$$

where θ_0 refers to output and θi to input.

This is not to suggest that such models are irrelevant. On the contrary his behaviour can be succinctly summarized by equations such as the original Tustin (1947) formulation

$$G(s) = e^{-ts} \left(As + B + \frac{C}{s} \right)$$

and the McRuer and Krendel (1959) summary

$$G(s) = \frac{Ke^{-ts}\,(T_L s + 1)}{(T_N s + 1)(T_I s + 1)}$$

The Tustin formula implies that the operator in a tracking task responds to the acceleration and velocity of the error as well as its magnitude. His relative attention to these three components is expressed by the size of the constants A, B and C. McRuer and Krendel suggest similarly that to describe tracking performance we need a lag time constant T_I of about ten seconds, a lead time constant T_L of about one second and a neuromuscular time lag T_N which is much smaller – about a tenth of a second, t is the reaction time. For more details of this approach see Meetham and Hudson (1969) McRuer and Weir (1969). Work of this kind does tend to be based on the performance of small numbers of subjects in highly restricted performance situations. It can be argued that any research must necessarily start in this way to form a basis of generalizations which can then be extrapolated to more complex behavioural situations. However, there are inherent difficulties:

1 Individual differences are not contained in these equations.

2 Changes due to learning, fatigue, motivation and instructions are also omitted.

3 It is assumed that the basis of human behaviour is the response to the immediate input. It can be argued that this is a fundamentally unsound assumption and that the essence of human behaviour is in the response to the total situation.

4 The operator response fluctuates apparently randomly, sometimes he overresponds, sometimes he underresponds, sometimes he waits, sometimes he responds very quickly. One variable within this context is the 'range effect' which describes the performance variation based on an average expected response. When the operator perceives an error of about the expected size he responds in an optimum fashion, if the error is less than average he overcorrects, if it is greater than average he undercorrects. Another variable is the so-called 'action threshold'. This can vary considerably; although there may be an average sized error about which he does something, the magnitude of this required to trigger a response is not consistent.

Machine dynamics

The behaviour of the machine in time is relatively predictable and immutable, it obeys the laws of physics which are consistent and are expressible in relatively simple equations. One important variable is the number of integrations which take place between a displacement of the control and the output of the system. In practice it is not quite as simple as this, for example a car accelerator with respect to the velocity of the vehicle is a mixed system involving no integration (the velocity), considerable output at one integration (the acceleration), and some at two integrations (the jerk); these vary with the speed with which the accelerator position is changed. Nevertheless the number of integrations is a crude summary of what happens and of the difficulty of controlling the vehicle – the greater the number of integrations the greater the stability and control problem. Thus a car has one, an aircraft two, a helicopter three and a submarine four. A diagram which summarizes submarine dynamics is shown in Figure 31. The helmsman has a control which looks just like the control-stick of an aircraft. What he does is to fly the submarine through the water controlling depth by changing the angle of the planes, which are like short wings on the sides of the submarine. Because of the massive power required his control actually changes the rate at which the plane angle changes, this eventually results in a change of plane angle which in turn eventually changes the pitch rate and then the pitch (the angle of the submarine to the horizontal). This finally results in a change in depth. Notice that from the operations point of view all these integrations feel like delays. Thus if he wishes to dive he pushes the stick forward, nothing will happen for some time; then the submarine will tilt and eventually the reading on the depth gauge will change. If he then decides he is near the bottom he will pull the stick back, but inexorably it will go on going down for some time before it responds again. A system like this is virtually impossible to control without aids to the operator in addition to extensive training.

The dynamic problem of interface design is the problem of fitting together these two systems, human and hardware, one obeying the principles of biology and the other the principles of physics. The operator can do quite a lot with his variable gain and his limited capacity to integrate and differentiate his inputs and outputs. However, the complexity of modern control systems is often such that he needs to be aided by deliberately manipulating the display and control dynamics to match the man and the machine together.

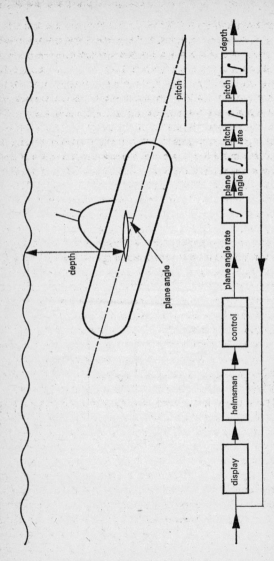

Figure 31 Control system for a submarine

Display and control dynamics

The techniques for manipulating interface dynamics are best illustrated by simple examples, although such examples may not fully convey the impact of these techniques on complex systems. Suppose we consider a car driver as a man changing the position of his vehicle along a road. This dynamic system is represented by Figure 32. It is not very easy to keep stable

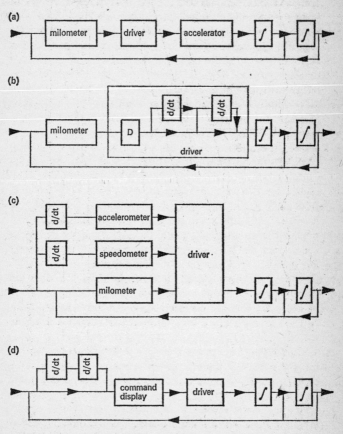

Figure 32 Manipulation of the display dynamics for the car driver

and smooth because the main control, the accelerator, essentially controls velocity although with an acceleration component as already mentioned. To keep the milometer reading increasing steadily, the driver must sense the velocity and acceleration of the vehicle. Actually he does this by the feel of the car movement and by looking through the windscreen, rather than by looking at the milometer but these are just additional cues which make his task easier. One can imagine how difficult it would be if his only cue to what was happening was the behaviour of the milometer. What he would then have to do for a smooth ride would be to detect the velocity and preferably also the acceleration of the milometer changes. This is shown in Figure 32(b). This is difficult, the driver as a dynamic system is now shown as a delay with two differentiations in series weighting his output (the gain factors which vary the weighting are omitted for simplicity). The task can be made much easier by, as it were, taking these differentiations out of the operator and building them into the hardware. It has already been noted that hardware is much superior for this kind of computation. The revised system with additional displays to aid the driver is shown in Figure 32(c). His task is now much easier. Another possibility remains: instead of giving his separate displays the designer can actually weight these data in some appropriate manner and then present them on one display as shown in Figure 32(d). The weighting would depend on the vehicle dynamics and the precise task required. Notice that the driver no longer knows exactly how many miles he has done or how fast he is going since these data are combined. In fact, of course, this is not done in cars because the task difficulty and variability do not warrant the cost involved. Instead, there is a compromise in Figure 32(c) and just a speedometer is supplied. It could be argued that a chauffeur-driven luxury vehicle should also have an accelerometer to increase the comfort of the passengers. However, the full aiding technique is used in aircraft for purposes of instrument guided glide path approaches. The instrument is called a zero reader as distinct from an ILS (Instrument Landing

System) indicator. An ILS indicator has two pointers at right angles. If the aircraft is on the glide path then both read zero in the centre of the dial. If the aircraft is off to the left then the vertical needle will be off to the left of zero, if the aircraft is too high the horizontal needle will be above zero. Suppose the vertical needle is on zero and the horizontal one is reading low, the aircraft is vertically below the glide path as shown in Figure 33(a). The pilot will take corrective action and will orientate his aircraft back towards the flight path. As soon as he is in the correct orientation to get back on to the optimum path the zero reader will read zero, but

I.L.S. indicator   zero reader

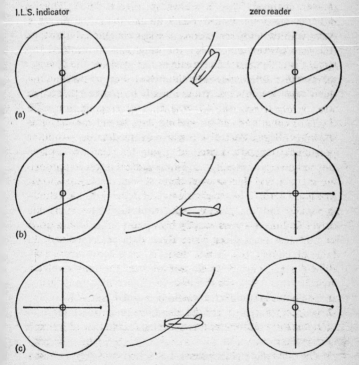

Figure 33 Display dynamics for aircraft landing

the ILS indicator will still read low (Figure 33b). If the aircraft then passes through the optimum path then the ILS will read zero but the zero reader will already read too high. Thus, given an ILS indicator, the pilot, unless he is very experienced, will oscillate above and below the optimum path; but given a zero reader he is always being encouraged or commanded to do the right thing. His task is simply to keep the zero reader reading zero. Notice, however, the pilot with an ILS indicator knows where he is although he is not being told what to do; the pilot with a zero reader is told what to do but although he knows he is doing the right thing he does not know exactly where he is. This technique is known variously as phase advance, display aiding, display augmentation or display quickening. Where the aiding is complete, as in the zero reader, then this is called a command display. *Quickening* is defined as any modification to a closed loop system which reduces the need for the operator to perform analogue differentiations. The complementary technique which reduces the need for human based integration is called *unburdening*. It is essential to ensure stability in complex dynamic systems with large time constants. Aided displays have the following advantages:

1 Training time is reduced.
2 Lower levels of operator skill or capacity are tolerable.
3 The operator has spare capacity to 'time share' with other jobs.
4 The effects of reversal errors are reduced.
5 It permits the introduction of built-in safety standards, e.g. maximum 'g' levels in aircraft can be built into the display dynamics.

There are a number of disadvantages:

1 The actual state of the system is less obvious.
2 In principle, if it can be done, there is no need for a human operator since it reduces his role to that of a simple amplifier.
3 The system is less flexible.
4 Dependence on the mechanism is increased.

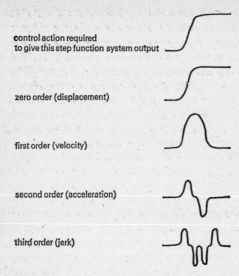

control action required
to give this step function system output

zero order (displacement)

first order (velocity)

second order (acceleration)

third order (jerk)

Figure 34 Order of control dynamics

For further details of this and related techniques see Kelley (1968) and Morgan *et al.* (1963).

It is also possible to manipulate the dynamics of controls in a corresponding fashion. This involves the concept of control order (Figure 34). A zero order or displacement control is one in which the system output is related to the control output by a constant (the gain factor). If the output of the control is integrated once to achieve the system output then this is called a first order control or velocity control. Two integrations take place in a second order or acceleration control, three in a third order or jerk control and so on. Broadly speaking, the lower the order of control, the simpler the task and the less skill is required; beyond second or at most third, order controls become impossible to use effectively even after extensive training. There are some situations in which although it increases complexity and required skill, it is nevertheless more efficient to use a first order rather than a zero order control, in general this is true in the traverse operation, for instance when a scan is being taken across a screen or a gun sight is being moved across the total field. The more the control is displaced the faster the controlled element moves. To reduce training problems and to maintain reasonable stability a mixed zero and first order system is often used; this is called a rate-aided control (Figure 35). An alternative, used in radar tracking is to have a variable order control, the control is zero order for precise positioning but

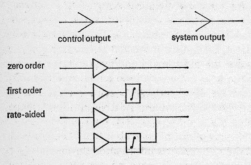

Figure 35 Rate-aided control

it can be changed to first order by pressing a button on top for rapid movements across the screen. The same problem can be dealt with by the rolling-ball control which has a very low gain but is very easy to move. Half a sphere several inches in diameter protrudes through a horizontal surface, it normally controls in two dimensions as does a control stick, but rapid controlled element movements can be made by hitting the ball tangentially, thus causing it to rotate quickly. Final precise positioning can be achieved by stopping it and making the appropriate small adjustments with the fingers in contact with the ball.

Display aspects of controls

Before an operator can use a control he must identify it and select it perhaps from a large array of other controls. This identification is easily achieved if the control is directly attached to the controlled member, for example the cross-slide control crank on a lathe. Failing this direct mechanical connection, the best form of identification is to keep the control physically adjacent to the controlled element or the representation of it, for example a pressure control valve directly beneath the appropriate pressure gauge. This problem of identification takes priority over unnecessary proximity. That is, it is usually more important to ensure that controls are easily identifiable than it is to put them all together; for example, an even array of identical switches in a car may look impressive at first sight, but the driver is likely to be perpetually making mistakes such as switching on the wipers when he needs the headlights. It is better to have the controls scattered all over the dash board and even to use different kinds of switches so as to avoid mistakes in identification. In summary, the ways of identifying control positions in relation to controlled elements are; relative positions, shape, colour and label.

A different display aspect of controls is that of action feedback. The operator always gets information through his tactile and kinaesthetic channels when he operates a control. This kind of information has two advantages over that received through the eyes and ears. Firstly it avoids the lags in machine dynamics: for instance an experienced driver will know from a changed pressure and position of the accelerator that it will eventually result in a given change in speed; this information arrives before the change in engine noise and the even later change in speedometer reading. Secondly this information channel is not dependent on attention and is relatively fast (tactile and kinaesthetic reaction times are slightly lower than visual or auditory reaction times: in changing gear on a car, if this is done by feedback control then not only does the driver not have to look at the gear lever, but additionally need not attend to this part of his activity). Thus there is always a strong case for ensuring that the operator receives appropriate action feedback. This factor lies behind the basic difference in performance when using pressure controls and displacement controls. A pressure control is one in which the response of the controlled element is a function of the force exerted on the control although the force does not result in an appreciable displacement. It has been argued that the basic output of the human operator is a force rather than a movement and thus such controls should reduce the output encoding which the operator has to do. This may be so, but such controls have a serious disadvantage in that there is no equilibrium position other than zero and any control action which must be maintained imposes a static muscular load on the operator. Thus, such controls are most useful where rapid brief control actions are required.

In a displacement control the response of the controlled member is a function of the distance moved by the control. There is usually some resistance to this movement and it is desirable that there should be to improve the action feedback. Displacement resistance can take many different forms:

1 Static friction in which there is force to be overcome to initiate a control movement. This is useful in reducing the possibility of accidental activation and also to reduce problems due to instability or vibration.

2 Sliding friction, also known as coulomb friction, which consistently opposes a control movement, it does not vary with displacement.

3 Elasticity – a force opposing a control displacement which increases with displacement. This is the kind of force provided by a return spring or for that matter by an elastic band.

4 Viscous damping – a force opposing movement which is proportional to the velocity of movement. The greater the speed of movement – normally in any direction – the greater the opposing force, usually provided by using a dashpot.

5 Inertial damping – a force opposing movement which is proportional to the acceleration. Provided simply by the mass of the control system.

6 Gating – sudden changes in the forces opposing movement at particular points in the displacement. Valuable in

providing cues about displacement. Found, for example, in well designed gear levers.

7 Powered damping – servo-control systems applied to control lever displacements. Characteristics can be deliberately designed for particular purposes which are often necessary to add 'feel' to power assisted control systems.

There have been various attempts to optimize control characteristics by quantifying the motor response of the human operator reacting with a control system which itself has a response which can be expressed as a dynamic equation (e.g. Chapanis *et al.*, 1949; Hick and Bates, 1950; Fitts, 1951), but it seems that there are too many contextual variables to allow of any general solution. It depends on the kinds of task, their relative importance (e.g. acquisition versus tracking), the skill of the operator, the relative importance of speed and errors and so on. Kelley (1968) points out that human operators cannot be described generally in terms of control theory in that there are differences in principle between human and automatic control systems. These differences originate in human consciousness, the deliberate choice of methods of achieving goals and the expression of these choices in movements.

Finally there are two aspects of control design affecting human performance which are essentially limitations of engineering design. These are 'backlash' and 'deadspace'. 'Backlash' is the tendency of a system to reverse when a control movement is stopped suddenly. It should be minimized by design, operator effects can if necessary be reduced by reducing control-display gain. 'Deadspace' is control movement which can take place without any system output. Again it should be minimized and any residue coped with if necessary by reducing control-display gain (McCormick, 1970).

Control-display relationships

The versatility of mechanical and more particularly electrical control systems is such that the relationship between the control actuation and the corresponding display movement may be unexpected. The expected relationships are often called 'population stereotypes'. This is useful jargon not only in identifying the phenomena but also in implicitly defining them. 'Stereotypes' implies habits and, in fact, the expected relationships are simply those which the operator is used to. 'Population' implies that they can be specific to particular populations of human operators rather than applying generally to all operators. This is not to suggest that these relationships are inherently superficial. On the contrary they can be very deep-seated and difficult to modify. The most basic stereotypes are extrapolations from cause–effect relationships based on the nature of the physical world and first identified by each individual in early childhood. For example, things that are pushed move away, things that are pulled move towards, in general things move in the direction of the applied force and movement is perceived essentially as relative to the individual. It follows that when an operator moves a control he expects the responding member and the display which represents it to move in the same direction as the control. For example, a lever which moves to the right should cause a pointer on a linear dial to move to the right. Moreover he perceives himself stationary and the moving members as moving relative to him. In this way a tool on a machine should move away from him when a lever is moved away from him as on the lathe. If the work moves and the tool is stationary as on a grinding machine then the work should move away from him when the control lever is moved away from him. There are less deeply embedded stereotypes associated with such mechanical devices as right-hand threads. For example, a clockwise rotation should cause the moving element to move away. There are also stereotypes associated with displays and controls separately. The first dial to be learned is the clock face which has twelve numbers increasing in a clockwise direction. Thus all readings are expected to increase clockwise and a clockwise control rotation is assumed to lead to an increase in the controlled variable, as in adjusting the volume on a radio.

The variation of stereotypes with population is at a more superficial level, for example the convention as to whether or not moving a lever down, switches power on or off. In general design problems of stereotypes are not difficult to deal with, the designer should follow his own intuitive feel for what is correct exercising particular caution if he is uncertain (which usually means that the user will be uncertain) or if he is designing for populations of which he is not himself a constituent.

Encoding

Stereotypes can be considered as a specialized kind of encoding topic in that the designer must be concerned with the relationship between the operator, the control and the world beyond the control as the operator perceives it. Encoding is essentially about perception and the models which the operator is using to facilitate the attachment of meaning to inputs. To appreciate this problem we have to go back again to the child learning to manipulate the world around it. The coordination of limb activities and sensory data eventually results in the acquisition of a spatial model of the surroundings. This process begins at birth and is well advanced by the time the child is two years old. There are fundamental implications; because the basic spatial model of the world was acquired so early it is regarded as natural and inevitable rather than learned, it is therefore extremely difficult to modify. At a later age the child gradually acquires the models associated with symbolic learning, that is conventional signs which indicate objects, ideas or processes. Stereotype behaviour is one indication of the existence of these perceptual models.

When the display is structured to provide compatibility with the operator's model then the operator has less encoding to do himself in updating his model using data presented on this display. Thus improvements of display encoding involve allocation of coding to the hardware side of the interface. Encoding, like dynamics design can be regarded as a sophisticated form of allocation of function (Singleton, 1967a).

It would be a mistake to take too far the analogy between man and machine-based informational encoding systems. There are unique characteristics in the way the human operator handles information as evidenced by the changes which occur in both learning and fatigue. The learning changes have been described in detail by Gibson (1968). Learning to perceive involves a long sequence of processes beginning with the appreciation of ranges of possible stimuli followed by the identification of inputs which appear to belong in sets in that there is co-variation. This leads to the isolation of external invariants which lie behind the changing inputs, e.g. the aircraft is of constant size and shape although it originates the retinal image which is continuously varying in shape, size and position as the aircraft moves. These external invariants must have their own meaning to the observer: what Gibson calls affordances, do they threaten him or satisfy one or more of his needs? Having identified external objects and systems the observer can begin to build up rules about how they interact with each other, that is cause-effect relationships, energy exchanges and information exchanges. Finally given his total model of what is happening around him the operator can attend selectively to those events which are likely to be important to him and he can ignore those which are not.

This description of the development in learning to perceive seems to apply usefully to the operator faced with an unfamiliar interface. He is likely to go through the above processes as he acquires skill in functioning within that particular man–machine system. If he gets into difficulties either because of unexpected events, because of the pressure of events or because he is tired then he is likely to regress down the same hierarchy of processes. That is, the first thing that goes wrong is a failure of efficient selective attention. Loss of control of inputs is evidenced by lack of certainty about what matters and what does not. This will be followed by suspicion that there is some undetected event behind the data which are no longer seen as related. The operator will then get concerned about what this means to him (the affordance) and he may have to regress to even more fundamental approaches, identifying a new external invariant or looking for new sets of co-varying inputs.

Some appreciation of the complexity of the activities of the skilled operator receiving information should at least persuade the display designer that his design task is never a straightforward one. The operator is essentially a decision maker and to make effective decisions he must have three kinds of information (Singleton, 1971b).

1 Information about policy and objectives. What is the operator trying to do? How much discretion has he? Within what limitations must he work?

2 Information about alternatives and consequences. The relative probability of achieving an outcome must be considered in terms of the relative desirability of each outcome and the relative consequences of not achieving it.

3 Information about the state of the system, what it is and how it is changing, together often with what it was and what it will be.

To summarize, display encoding for the human operator is an art which depends on understanding what he needs to know and putting the information in a form which matches the model of the situation which he is using. This model may change as the skill develops or even within the conduct of a particular task, it can also be changed by the actual method used to encode the information. It follows that the well coded display is not necessarily the easiest to use on first acquaintance. It can be designed to encourage the operator to develop the optimum model.

The various categories which may need to be combined in the total display structure are shown in Table 18. Stimulus uncertainty and response uncertainty require information respectively on what is happening and what to do about it. Irrelevancy is not necessarily to be reduced, it may provide realism – a coloured picture compared with a black and white picture – and it may help to keep up the arousal level.

Redundancy also is not necessarily undesirable; it can reduce the effects of noise and it can often provide reassurance – most of the conversation between a pilot and a ground controller is redundant as information but it provides mutual confidence and awareness. Current information is obviously

Table 18 **Kinds of information within a display**

Stimulus Uncertainty
Response Uncertainty
Relevant – Irrelevant
Redundant
Past – Current – Future
Anticipatory

necessary but so also may be past and future information, the past often helps in anticipating the future but there are also data which are essentially anticipatory, for example rates of change.

Psychophysics

Matching displays to the characteristics of the sense organs is essentially a problem of achieving an adequate signal to noise ratio.

In the case of vision the signal can be strengthened by increase of size, brightness, contrast and time available. Keeping down the noise level begins by reducing clutter of displays or displayed data. Thereafter it is a matter of reducing glare. This involves keeping light sources at least 60° away from data sources or providing shields. Low light source intensities provide a general form of insurance against glare. This can be done either by using many light sources or by using fluorescent rather than filament sources. This is not to suggest that uniform lighting is desirable, on the contrary some variations in illumination levels are essential to stimulate arousal and to improve the structure of the visual field. Indirect glare reflected from dials, interface surfaces and so on should also be avoided. In the case of hearing, communication is improved not only by keeping signal levels up and noise levels down but also by redundancy. This is particularly true of speech. In the case of kinaesthesis the improvement of signal strength has already been discussed in relation to controls. Noise in this context comes from vibrations which should be kept to a minimum for this as well as for other reasons.

Design procedure

The general approach to interface design is summarized in Table 19. The starting point of design is the characteristics

Table 19 **General approach to interface design**

Type of operator	Programmed – concept
Function	Setting-up, operation, maintenance
Interface	Encoding
	Dynamics
	Grouping: function
	sequence
	priorities
	Stereotypes
	Anatomy
	Psychophysics

of the operator and what he wishes to achieve. These can be considered under the headings of type of operator and function. Whether the operator is programmed in the sense of generating a learned response to a given stimulus or a concept operator in the sense of being updated and left to generate his own optimum solution will obviously have quite fundamental influences on the interface design. Within these dimensions also comes the problem of identifying the kind of model which the operator is using. The particular function – setting-up, operation or maintenance – will also effect the interface. In fact most interfaces have a mixed function in this sense but nevertheless the purpose of a particular interface element can be considered from this point of view. Finally in considering the interface itself the two main sets of problems can be divided – as in this chapter – into encoding and dynamics. Grouping of interface elements by function, sequence or priorities is a specialized aspect of the encoding problem. Stereotypes are a set of problems having both encoding and dynamics aspects. Finally there are always problems of functional anatomy and psychophysics.

This approach presupposes that it is possible and desirable in the particular design case, to proceed systematically from general principles of human behaviour. This is not necessarily true, there are other possible approaches (Singleton, 1969). Intuitive display design is often justifiable on grounds of speed and economy and can be very effective providing that the designer has user experience and providing a checklist is used to confirm that there are no obvious oversights (Table 20). Another approach, often favoured by engineers, is to start from what the hardware does. This is at least easier to identify than what the human operator does. Information is displayed about everything that is happening, or rather everything that can be detected and instrumented, and controls are provided for whatever variations are possible within the machine or process. The displays and controls are then presented for use in some commonsense manner and are modified or eliminated by experience. These three approaches – engineering, intuitive and behavioural – are not, of course, mutually exclusive. The choice of procedure or relative weighting of various procedures depends on the particular case. The basic premise which is self-evidently valid is that man–machine interfaces must take account of human advantages and limitations (Singleton, 1971c).

Table 20 Display design checklist (Singleton, 1969)

Dimension	General problem	Possible objectives	Parameters, techniques and principles
Necessity	Does the user's need justify the provision?	Is it worth the expense and complication? Can the operator manage as well or better without it?	Cost-effectiveness studies Justification other than on grounds of information: tradition apparent quality styling
		What happens if it malfunctions?	Critical incident studies Possibilities: presents no information presents false information
Sufficiency	What data may the operator need which have not been provided?	Would more data be useful: as routine? to set up or change the system? in case of malfunction? about past state? present state? future state?	Man-machine allocation of function Task analysis use of: storage displays quickened displays predictor displays

Table 20 Display design checklist (Singleton, 1969) – continued

Dimension	General problem	Possible objectives	Parameters, techniques and principles
Legibility	Can an operator with average senses see or hear what is required easily?	Is unnecessary activity required to extract the data?	Activity may: take longer time involve more concentration effort involve change of posture position
		Is the presentation about the right strength?	Visual parameters: position size orientation illumination pointer size shape contrast lettering size style excess scale division graduation clutter

Auditory parameters:
amplitude
frequency
attack
clarity

Lighting parameters:
contrast
shadow
glare
reflections

Sound parameters:
background noise

Connections with real events:
physical
symbolic
legends

Error summation studies

Is the signal–noise ratio high enough?

Is it clear where the data are available?
Is it clear what the presented data are about?

Have unnecessary data been eliminated?
Is the accuracy appropriate in terms of:
instrumentation?
perception?

Table 20 Display design checklist (Singleton, 1969) – *continued*

Dimension	General problem	Possible objectives	Parameters, techniques and principles
Compatibility	Does the display conform: to the real world?	Is the operator using real – artificial displays? If so: does movement of indicators correspond to real movements? Is the pattern of display items similar to real things or data sources?	
	to other display items?	Have the most appropriate grouping principles been followed?	Possible grouping principles: sequence of scanning priority frequency of use similarity or identity of function
		Have the most appropriate additional grouping cues been used?	Possible grouping cues: colour shape size position
		Are different dials, pointers consistent when appropriate?	scale style pointer style

with previous habits and skills?	Will the operator require special training?	Skills analysis
	Is there likely to be a discontinuity with earlier operator experience?	Population stereotypes
with required decisions and actions?	Does the presentation encourage the operator to: Think about the right problem in the right way? Select the right action and move in the right direction?	Encoding to match perceptual models Control–display relationships

8 Particular Kinds of Systems

The picture of a typical system which emerges from this book may well be of some complex powerful mechanism controlled by a skilled human operator, the two being linked by an elaborate interface with many different controls and even more dials and other displays. Examples are a pilot flying an aircraft, an engineer controlling a power station, a shift worker in an oil refinery, or an operator working with a large machine-tool. It is true that these are typical of the kinds of man–machine system where this design philosophy has been used and where in fact it increasingly proves essential for the creation of efficient systems. Examples have been quoted from such systems because they exist and because these are the kinds of system where the approach is most obviously useful. Nevertheless the approach is much wider than this and it is of sufficient generality to provide insights into any kind of human situation where there is a goal to be achieved and the people involved have various aids, designed by others, to facilitate their progress towards this goal. Educational technology is a problem of allocation of function between teachers and hardware, and an interface design problem between hardware, teachers and pupils. The design of kitchens is a systems problem involving hardware and the housewife, with all the problems of allocation of function, interface design, job aids design and training. Any office has problems of what clerks do, of what computers and other machines do, how the whole system is controlled, the pattern of information inputs, outputs and processing and so on. The clinical diagnostician is an operator in a complex system with problems of integrating many streams of information, storing information and making decisions which involve elaborate trade-offs of potential achievements with success and consequence of failure. Various kinds of systems have been selected for discussion in this chapter, partly on the grounds of their frequency of occurrence, their importance to the community and as wide a sample as possible of all system designs.

Production systems

A production system requires the integration of numerous specialist activities with the aim of generating a particular range of products. There are three different approaches to the philosophy of organization (O'Shaughnessy, 1966).

The traditional method is to allocate activities and duties by social class. That is, members of particular families, those with capital and those with a particular exclusive education are accepted as the leaders of the organization. They control and direct activities except at the lowest levels. The objectives are financial and sometimes paternal. The incentives to work are essentially punishments (fear of dismissal) rather than rewards. A human relations approach emerges where there is full employment, strong trade unions and stable industrial organizations with little competition. It is considered that although disciplinary and financial incentives may be necessary they are not sufficient. Considerable attention is paid to motivational problems, particularly those stemming from the behaviour of groups. The objectives are stated in terms much wider than merely making money, but they are usually correspondingly vague. There is talk of service to the community generally as well as to employees.

The third approach is that based on systems theory. The organization is regarded as self-adaptive with the basic goal of survival. In this sense shareholders, management and workers share a common goal. Functions are achieved by subsystems linked by information channels. Objectives can be stated in relation to four groups: shareholders, managers, workers and consumers. The shareholders require profits, the managers require a career structure, the workers require financial and welfare incentives, the consumers require a service or a product. The separation of functions depends on the product but there is one general problem: the degree of autonomy of local organizations. The original autonomy principle of fifty years ago or more was necessary because of

Table 21 **Relative advantages of real and artificial displays**

Real	Artificial
Contain variety of unquantifiable evidence	Essentially quantitative

Essentially in real time	Information can be about:
	    past
	    present
	    future
	Can be:
	    anticipatory
	    time compressed
	    time expanded
	    differentiated
	    integrated

Usually the more reliable	Subject to failure of sensors and computers
Observer can fully exercise his selecting and structuring abilities Maximum versatility	Selecting and structuring done by designer often easier to use but less versatile
Usually the less costly	Can be expensive
Difficult and expensive to record in detail	Easy to record

poor communications systems; improvements in communications led to the centralization of decision making, sometimes excessively. Realization of the importance of real displays as the context for artificial displays has caused some reversal of the centralizing tendency. In more colloquial terms formal data are never adequate, the decision maker always needs to go and see for himself (Table 21). The man–machine allocation of function problem within management is currently to do with computers. Excessive enthusiasm for the computer has led to many distortions of allocation and even to the concept of the manager as equivalent to a computer. In this way we have the jargon that

Table 22 Relative advantages of clerks and computers

Clerks	*Computers*
Superior error detection procedure storage multiple access to store material	Superior speed accuracy literal storage
Can detect dishonesty	Free from emotional responses laziness dishonesty
High tolerance of ambiguity vagueness uncertainty	Requires precise instructions and precise evidence
Easy to programme	Tedious to programme
Intelligent therefore versatile	Encourages fixed techniques
Available in large numbers	Still relatively rare
Relatively cheap but cost per unit increasing	Very expensive but cost per unit of computational power decreasing rapidly

electronic machines are 'dry computers' and people are 'wet computers'. While restressing all the provisos already discussed about comparing people and machines it is still useful to contrast the relative performance in the management context (Table 22). The difficulties of electronic data processing have been accepted theoretically for a decade or more and are now being justified in practice by the poor performance of many automated systems. These fundamental problems are:

1 Because computers will only accept numbers there is a

tendency to use data based on ease of measurement rather than relevance.

2 As a consequence factors that can be easily measured tend to be overweighted in decision-making.

3 Speed of introduction is often achieved by the mechanization of existing procedures. This is almost certainly not the optimum solution, new techniques require new procedures.

4 Good new computer based systems are compared with poor clerically based systems to the advantage of the former. Often it might be better to make an investment in improving the clerical system particularly the personnel; this would be cheaper than computerization.

5 The ease of data handling within computers encourages excessive growth in information flow.

6 Information handling is given too much attention relative to information presentation to the decision-makers.

7 The dominance of the information handling system leads insidiously to a growth in middle management personnel many with merely technical skills. There is a relative reduction in numbers and assumed importance of supervisory staff in direct contact with the production workers.

The problems of presentation of information to managers are not all that different from problems of presentation to machine operatives. The information flow system is analogous (Figure 28). There is more reliance on static displays (often computer outputs) rather than dynamic displays such as dials and the manager usually has more variety of strategy so that his programme is provided more in terms of ends rather than means. Broadly, in moving up the status hierarchy in industry there is less reliance on real displays, more filtering and summarizing of data, increased use of graphical data and less direct definition of procedures and responsibilities. Nevertheless all levels of management need a programme and the ways in which this is formulated and provided are worthy of study.

Usually there is a long term programme which is based on or takes the form of the company policy. Then there are more short-term objectives to do with production schedules. The manager obviously needs to know what the programme is and how successfully he is meeting it. These factors determine the design of displays for him. There is a current fashion (management by objectives) to allow managers to set their own objectives within company policy. This provides high incentives not only because individuals feel more in control of their own affairs but also because, in common with all workers, they are prone to overestimate what can be achieved within, say, the next year. They therefore set themselves over-optimistic objectives and have to work very hard to attempt to justify them. Thus, there are problems of objectives, allocation of function, interface design and training design.

Educational systems

From this point of view education is not qualitatively different from training and can, in fact, be regarded as a specialized kind of training (Singleton, 1968b). It fits within the system design context as shown in Table 23. In common

Table 23 The educational system as a training system

Training	*Education*
To fit operator within a particular system	The community is the system
Worth evaluated in terms of contribution to system output	Contribution to gross national product and national way of life
Financed by parent system	Government represents parent system

with all training, education requires subsidy by the parent system – in this case the state. The objectives are to maintain the standard of living of the community by providing personnel who will ensure the efficient supply of goods and services and also to maintain the quality of life. This latter gets more and more difficult to define as the standard of living rises. Obviously if the basic necessities of food and shelter are not adequate these must be dealt with before other needs are considered. Maslow (1963) suggested a hierarchy of needs, beginning with those to do with physiology and safety and working up to esteem and self-actualization. The assessment of the relative importance of these higher needs is clearly a matter of opinion and educational discussion is often confounded by disagreements in this area.

The separation of functions in the sense of various educational sub-systems (primary, secondary, tertiary, academic, technical, commercial, artistic and so on) is con-

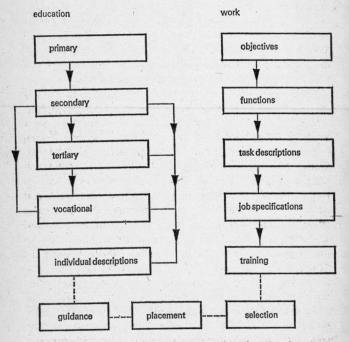

Figure 36 The matching problem between the educational system and the work system

founded not only by disagreements about needs but also by ignorance about the future. There is no way of estimating accurately how many doctors, civil engineers, chemists, salesmen, clerks, machine tool operators or labourers, a community will need in twenty years' time. There is therefore no way of designing the system in terms of its output. We have no formal evidence from which to calculate what the school-leaving age should be, how many university places are required and so on except on the basis of input data. We do know about birth rates, we can estimate how many children will achieve a given educational standard. Input only controls with a twenty year time lag are, of course, very dangerous in that the system could go wildly unstable. Indeed it appears to be doing just this in most advanced countries at the present time. The remedy can only come from more systematic man-power planning at the national and regional levels but this depends on improvements in Task Analysis, Job Analysis and Skills Analysis.

The man–machine allocation of functions problem and interface design problems involve what is now called educational technology. The allocation of function criteria (Chapter 2) apply, so do all the principles of interface design and job aids design. Language laboratories, teaching machines and simulators need to be developed and utilized on systems design principles (Chapters 5 and 6). There is another allocation of function problem to do with kinds and levels of individuals involved in teaching. This is what is called, in other jargon, the business of team teaching and teaching auxiliaries. The danger in education is the theorist with a mission. He selects the evidence which supports his strong prejudices, problems are discussed and sometimes solutions are imposed in such widespread and general terms that there can be no rational discussion. One remedy is the systems design approach which will at least ensure that problems are separated into comprehensible groups and that relevant evidence is searched for and gaps in it are recognized.

Military systems

These systems problems stimulated most of the original thinking about design. This is partly because there are large numbers of supporting research institutes which have time and other resources to devote to thinking about these problems and also because there is an obvious need to get the benefits of advancing technology as rapidly as possible. There are some corresponding liabilities in that most of the scientific support services are based on engineering and the natural sciences. Often specialists from these fields are more difficult to convince about the importance of the human operator than are non-scientists. The formal command systems are also an obstacle, it is difficult to accept or to use a philosophy which assumes that hardware design and training are complementary aspects of one problem if the designers and trainers have no communication except remotely through the users.

On the other hand there is a general military tradition of respect for the quality of the individuals or groups of operators. For example, a commanding officer is usually more concerned about the state of morale among his men than a managing director is about the morale of his workforce. The training problem is accepted as important if only because for a weapon system in peacetime, the only use is during training. The objectives of a system tend to be very simple in wartime, but very complex in peacetime. The design of a new tank for example is complicated by the need to provide wide versatility, if it were known that it would be used only in Northern Europe the design problems would be much simpler. The possibility must be considered that it might be utilized in mountainous country, in cities to quell riots, in very cold or in very hot conditions. Allocation of function principles based on existing personnel must also take account of the fact that, in wartime, the personnel characteristics would change very quickly as new recruits became available, all untrained but some of very high quality. The interface designs must be relatively simple not only because of rapidly changing personnel but also to take account of possible reduction in skill of operation under stress conditions. Problems vary from designing digging tools and tanks to designing navigational aids and vast early warning detection systems. There are complicated communication problems between specifiers, designers, evaluators, trainers and users where the common language provided by the systems approach can be invaluable if it is generally acceptable.

Military problems remain the main source of new ideas. New problems include, for example, how to design hardware and information systems for a training role as well as an on-line role, the optimum relationship between on-line operators and maintainers, between makers and users and between sub-system operators and system operators, the effects of stress particularly combinations of stresses such as heat, vibration and danger, continuous activity over periods of twenty-four hours or more and so on. All of these will no doubt eventually have 'spin-offs' into the industrial and other environments.

Environmental systems

Surroundings are important to the person at many different levels. The basic level is that he wishes to survive without damage, and thus excessive heat or cold, noise, vibration and acceleration must be avoided. These problems are not quite as easy to deal with scientifically as they might appear at first sight. The difficulty is that the dimensions found to be simple in physical science terms are not so simple in biological or behavioural terms and vice versa. For example, the electromagnetic spectrum makes for a nice integrating concept of all radiation but it does not provide a method of predicting or measuring effects on people: different parts of the spectrum provide quite different sensations. Conversely the unitary human concept of feeling cold for example is affected by a number of physical science variables such as temperature, humidity and windspeed. Thus for each potential environmental stressor a complex index must be devised which incorporates a variety of physical variables. It is then necessary to evaluate several risk levels such as immediate danger, longer term danger, performance decrement and discomfort. All this is complicated enough and the situation could hardly be described as fully sorted out for any of the above variables; certainly we know almost nothing about the combined effects of stressors (Poulton, 1970).

Things get even more complex if we begin to look for mild stress effects. What happens to the learning efficiency of school children if the class room is slightly too hot or too cold? What happens to travellers subjected to moderate noise and vibration levels? What are the effects on pedestrians of exhaust fumes in city streets? There is some research in this area (Davis, 1970) and progress is fairly rapid, but it is not adequate to supply comprehensive evidence to facilitate decisions where there are obvious cost penalties of proposed improvements.

Beyond the mild stress problem we come to the concept of amenity. In this area it is not yet clear whether or not the dialogue should be conducted in cost terms. It can be argued that, to make a reasonable and rapid impact on the planners and other policy makers, it is necessary to translate the problem into financial terms so that comparison with other relevant factors is direct (Waller 1970). It can also be argued that, to take a games analogy, it is always easier to win when playing at home rather than playing away, and that the research worker concerned with higher level human functions and aspirations should refuse to conduct the argument at a mere financial level. On the one hand it does not seem entirely appropriate to assess the value of a cathedral in terms of what people are prepared to pay to preserve it, or to measure the effect of noise from an airport in terms of the reduced market value of houses under the flight path. On the other hand it is hardly legitimate for the environmentalist to campaign that something must be done without specifying who must do it and who must pay for it. The danger in this area of study is that the investigators become men with a mission so that, either deliberately or unconsciously, they generate, select and publicize evidence that supports their own prejudices. Even worse is to utilize the status of science and technology to support a claim for which the protagonist's opinion is no more and no less important than that of a layman.

The remedy is to stick to the evidence and to methods of generating evidence which control individual bias. It must be accepted that this leaves important areas of human decision making, at least temporarily, without the support of systematic inquiry. If there is no way of generating valid evidence then there is nothing to be said by the technologist, however important the problem is in human terms. Many problems of educational methods, of violence in society, of penology and of drugs come in this category. Inevitably the higher level of the problem the less secure is the evidence and the less systematic are the design procedures. Consider for example an aircraft passenger compartment. There are physical environment problems of air supply, noise and so on. These can be measured, effects predicted and appropriate action taken by the designer. There are functional environment problems of mobility, provision of facilities, supply of food and drink and so on. These also can be dealt with in a satisfactory fashion making a reasonable trade-off between the system as a flying machine and the system as a passenger carrier. There are then problems of comfort, pleasure, security, personal space and other high level amenity variables. Here the system designer can only consult with industrial designers and interior decorators who work as artists rather than scientists. Consider the problem of office design; how is a decision to be made about whether to incorporate open offices or closed offices? How can evidence be accumulated relevant to this decision? It is not very much use asking the people who will work there. It has been demonstrated that the opinions of those who have worked in open plan offices are very different from those who have no experience but are asked if they would like to. There are isolated bits of reliable evidence, for example Canter (1968) showed in one experiment that test performance was poorer in larger offices, but how valid is this in relation to the overall objectives of persons working in such offices and of those employing them?

The present state of knowledge in this field is such that research on techniques and principles is needed even more than evidence on particular cases. This is not to deny that the techniques and principles might best be generated by studies of particular cases. Given that the answer to a question is unknown or even currently unknowable the proper strategy is to leave adequate flexibility in the system for modification based on experience. For example, rather than attempt to predict pedestrian flows between a group of buildings, the whole area can be covered in grass, and paved ways can be added later following the paths worn through the grass. This kind of contingency reserve in design is usually obtained by allocating functions to human operators and adding further procedures and hardware later as it proves desirable.

Glossary

ability A human resource within an individual which has been developed by education and training.

action threshold The size of error required to elicit a correcting action. Can vary considerably within and between individuals.

algorithm A set of rules for solving a problem in discrete steps.

allocation of function The business of deciding how the total functions required to achieve some purpose can best be divided – by implication between man and machine or between men, machines and procedures.

alpha numeric symbols The letters of the alphabet (26 or 52) and the numbers (10).

analogue computer A computing system in which the magnitude of signals represents the information.

aptitude Special capacity to acquire particular kinds of skill.

attitude A predisposition to behave in a certain way. May be a personality factor or a consequence of instructions.

backlash Characteristic whereby a controlled member overshoots in the opposite direction when a control movement is terminated.

black box A system about which only inputs and outputs are knowable.

capacity A potential ability or a human resource which is available but not subject to training.

computer Mechanism predominantly concerned with information handling. Sometimes subdivided into dry-computers (hardware) and wet-computers (human operators). See analogue computer and digital computer.

concept operator One who makes his own decision about the appropriate response from an understanding of the situation triggered by the stimulus, contrast with programmed operator.

coulomb restoring force The same as sliding friction.

data Signals to human operators.

deadspace Region of movement in which a control movement has no effect. The hardware equivalent of an action threshold.

decision A choice made between alternative courses of action.

digital computer A computing system in which the signals which represent the information are made up of digits.

displacement control One in which the controlled variable is a function (not necessarily linear) of the distance moved or the angle turned through by the control.

dynamics Time relationships or effects dominated by the time variable, notably timing of human responses. Hence relevance to control design.

elasticity Control characteristic such that force opposing movement increases with extent of movement.

element Part of a task which can be postulated as occupying real time, e.g. a movement or a visual scan.

encoding – by implication of display data. The design of the structure of presented information to match the inferred perceptual processes of the operator.

ergonomics Systems design with the characteristics of the human operator as the frame of reference.

function Activity which achieves or aids in the achievement of some objective.

gain The relationship of input to output. Applied to control – display interaction or control – controlled member effect or effect of human operator. Expressed as a ratio output over input.

GOG 'Getting off the ground' – the set of functions concerned with setting up a system.

hardware System components made of inanimate material, contrast with software.

inertia A force opposing a control movement which is proportional to the acceleration of movement, that is, to the force generating the movement.

information Data which reduce uncertainty.

interface Imaginary plane across which information and energy are exchanged – by implication between man and machine.

job Set of tasks which one operator is employed to carry out.

job aid An off-line display which presents data relevant to the task. Normally static and used as a memory aid, e.g. a legend plate on a machine tool or an instruction manual.

language A conventional set of symbols and symbol combinations sometimes with related utterances which is used to convey information between members of a community.

liveware Human operators, described in this way to highlight the potential range of division of functions between hardware, software and liveware.

machine Mechanism designed to extend human functions.

need Human condition prompting action and explained by the lack of something.

occupation A kind of employment recognized as serving some need of the community, made up of sets of tasks which are not necessarily identical for all members of the occupation.

off-line An activity is on-line or off-line depending on whether the system is or is not changing at that time and place, i.e. functioning to achieve the purpose.

on-line See off-line.

order (of controls) The specification of the function relating control movement to controlled member movement. In zero order or position control the relation is linear. In first order or velocity control, the control movement is integrated once. In second order or acceleration control two integrations are involved. In third order or jerk controls three integrations are involved.

perception The process of attaching meaning to sensations.

pressure control One in which the controlled variable is a function (not necessarily linear) of the force exerted on the control which moves negligibly.

prime equipment The hardware that functions on-line as distinct from maintenance or training equipment.

programmed operator Also called stimulus-response (S-R) operator. One who makes an instructed (programmed) response to a given stimulus. For successful training all stimuli and all stimulus-response combinations must be predictable. Contrast with concept operator.

psychophysics The relationship between physical stimuli and sensory processes. Relevant to the problem of transmitting information from a machine to a man – hence display psychophysics.

purpose Intention to satisfy a need.

quickening A modification to a closed loop system which reduces the need for the operator to perform analogue differentiations.

range effect The effect whereby a skilled operator makes an optimally damped response to an average sized error, an overdamped response to an unusually large error and an underdamped response to an unusually small error.

S-R operator See programmed operator.

sensation The experience dependent upon a stimulus affecting a particular kind of receptor such as eyes or ears.

signals Physical representation of an event.

sliding friction A force which opposes a control movement in any direction.

software Set of routines which can modify the system performance without modifying the hardware structure, contrast with hardware.

static friction A force which opposes a control displacement at the initiation only, that is, it becomes zero when the control is moving.

symbol A written sign as opposed to a verbal one which by convention represents something (the referent).

system Set of related objects.
Natural systems – not designed and with no objectives.
Man-made systems – designed to serve human needs as described by the objectives.

task Function allocated to a human operator rather than to a machine.

time sharing Situation in which an operator is simultaneously engaged in more than one task.

training Procedure for the encouragement of learning.

training aid An off-line display which presents data relevant to the building up of skill required for a particular task.

unburdening A modification to a closed loop system which reduces the need for the operator to perform analogue integrations.

viscous damping A force opposing a control movement which is proportional to the velocity of movement.

white box A system for which functions are separable.

References

ANNETT, J. (1969), *Feedback and Human Behaviour*, Penguin.

ANNETT, J. (1971), 'Learning in practice' in Warr, P. B. (ed.), *Psychology At Work*, Penguin.

BAINBRIDGE, L., BEISHON, R. J., HEMMING, J. H., and SPLAIN, M. (1968), 'A study of real time human decision making using a plant simulator', *Operational Research Quarterly*, vol. 19, pp. 91–106.

BARTLETT, F. C. (1943), 'Fatigue following highly skilled work', (Ferrier lecture) *Proc. Roy. Soc.* B 131, pp. 248–57, Reprinted in Legge, D. (ed.), 1970, *Skills*, Penguin.

BEISHON, R. J. (1967), 'Problems of task description in process control' in Singleton, W. T., Easterby, R., and Whitfield, D. (eds.), *The Human Operator in Complex Systems*, Taylor & Francis.

BIEL, W. C. (1962), 'Training programs and devices' in Gagné, R. M. (ed.), *Psychological Principles in System Development*, Holt, Rinehart & Winston.

BOWEN, H. M. (1967), 'The imp in the system' in Singleton, W. T., Easterby, R., and Whitfield, D. (eds.), *The Human Operator in Complex Systems*, Taylor & Francis.

BROWN, J. A. C. (1954), *The Social Psychology of Industry*, Penguin.

CANTER, D. (1968), 'The psychological study of office size: an example of psychological research in architecture.' *Architects Journal*, March.

CHADWICK-JONES, J. C. (1969), *Automation and Behavior – Individual and Social*, Wiley.

CHAPANIS, A. (1959), *Research Techniques in Human Engineering*, Johns Hopkins.

CHAPANIS, A. (1960), 'Human engineering' in Flagle, C. D., Huggins, W. H., and Day, R. H. *Operations Research and Systems Engineering*, Johns Hopkins.

CHAPANIS, A., GARNER, W. R., and MORGAN, C. T. (1949), *Applied Experimental Psychology*, Wiley.

CRAWFORD, M. P. (1962), 'Concepts of training' in Gagné, R. M. (ed.), *Psychological Principles in System Development*, Holt, Rinehart & Winston.

CRESSWELL, C. W., and FROGGATT, P. (1963), *The Causation of Bus-Driver Accidents*, Oxford University Press.

CROSSMAN, E. R. F. W. (1956), 'Perceptual activity in manual work', *Research*, vol. 9, pp. 42–9.

CURRIE, R. M. (1960), *Work Study*, Pitman.

DAVIS, P. R. (1970), (ed.), *Performance Under Sub-optimal Conditions*, Taylor & Francis.

DEGREENE, K. B. (1970), 'Systems analysis techniques' in DEGREENE, K. B. (ed.), *Systems Psychology*, McGraw-Hill.

DEP (1967), 'Glossary of training terms', *Department of Employment and Productivity*, HMSO.

DUNN, J. (1971), 'Skills analysis for outdoor work involving mobile powered tools', *Final Report to Medical Research Council, London*, September 1971

EASTERBY, R. S. (1966), 'Design of a lathe for international markets.' *Human Factors*, vol. 8, pp. 327–38.

EASTERBY, R. S. (1967a), 'The grammar of sign systems', *Print*, vol. 13, pp. 6.

EASTERBY, R. S. (1967b), 'Perceptual organization in static displays for man-machine systems' in Singleton, W. T., Easterby, R., and Whitfield, D. (eds.), *The Human Operator in Complex Systems*, Taylor & Francis.

EDHOLM, O. G. (1967), *The Biology of Work*, Weidenfeld & Nicolson.

EDWARDS, E. (1971), 'Techniques for the evaluation of human performance' in Singleton, W. T., Fox, J. G., and Whitfield, D. (eds.), *Measurement of Man at Work*, Taylor & Francis.

ELY, J. H., BOWEN, H. M., and ORLANSKY, J. (1957), 'Man–machine dynamics' quoted by Chapanis, A. (1960), 'Human Engineering' in Flagle, C. D., Huggins, W. H., and Roy, R. H. (eds.), *Operations Research and Systems Engineering*, Johns Hopkins.

EMERY, F. E. (1969), *Systems Thinking*, Penguin.

FERRELL, W. R., and SHERIDAN, T. B. (1967), 'Supervisory control of remote manipulation', *Institute of Electronic and Electrical Engineers Spectrum*, vol. 4, pp. 81–8.

FERRIES, G. W. (1968), 'The first steps in training design' in Robinson, J., and Barnes, N. (eds.), *New Media and Methods in Industrial Training*, BBC.

FITTS, P. M. (1951a), (ed.), *Human Engineering for an Effective Air-Navigation and Traffic-Control System*, National Research Council.

FITTS, P. M. (1951b), 'Engineering psychology and equipment design' in Stevens, S. S. (ed.), *Handbook of Experimental Psychology*, Wiley.

FITTS, P. M., and JONES, R. E. (1947), 1. 'Analysis of factors contributing to 460 "pilot-error" experiences in operating aircraft controls'; 2. 'Analysis of 270 "pilot-error" experiences in reading

and interpreting aircraft instruments', Reprinted in Sinaiko, H. W.
(ed.) (1961), *Selected Papers on Human Factors in the Design and
Use of Control Systems*, Dover Publications.

FITTS, P. M. (1964), 'Skill learning' in Melton, A. W. (ed.),
Categories of Human Learning, Academic Press.

FLEISHMAN, E. A. (1966), 'Individual differences' in Bilodeau, E. A.
(ed.), *Acquisition of Skill*, Academic Press.

GAGNÉ, R. M. (ed.) (1965), *Psychological Principles in System
Development*, Holt, Rinehart & Winston.

GARNER, W. R. (1962), *Uncertainty and Structure as Psychological
Concepts*, Wiley.

GIBSON, J. J. (1968), *The Senses Considered as Perceptual Systems*,
Allen & Unwin.

HICK, M. E., and BATES, J. A. V. (1950), 'The human operator of
control mechanisms', *Ministry of Supply*, Monograph, vol. 17,
p. 204.

HOBBS, J. A. (1967), 'The work of the road accident injury group',
Road Research Laboratory, Ministry of Transport,
RRL report LR 108.

HORNE, J. (1969), (ed.), 'A study of the initial training for firemen',
A.P. Report 30, Applied Psychology Department,
University of Aston in Birmingham.

JONES, J. C. (1970), *Design Methods*, Wiley.

JORDAN, N. (1963), 'Allocation of functions between man and
machine in automated systems', *Journal of Applied Psychology*,
vol. 47, pp. 161–5.

KELLERMANN, F. Th., van WELY, P. A., and WILLEMS, P. J.
(1963), *Vademecum Ergonomics in Industry – Philips Technical
Library*, Cleaver-Hume Press.

KELLEY, C. R. (1968), *Manual and Automatic Control*, Wiley.

KIDD, J. S. (1962), 'Human tasks and equipment design' in
Gagné, R. M. (ed.), *Psychological Principles in System Development*
Holt, Rinehart & Winston.

MCCORMICK, E. J. (1970), *Human Factors Engineering*,
McGraw–Hill.

MCRUER, D. T., and KRENDEL, E. S. (1959), 'The human operator
as a servo-system element', *Journal of the Franklin Institute*,
vol. 267, part 5, pp. 381–403, part 6, pp. 511–36.

MCRUER, D., and WEIR, D. H. (1969), 'Theory of manual vehicle
control', *Ergonomics*, vol. 12, pp. 599–633.

MASLOW, A. H. (1943), 'A theory of human motivation',
Psychological Review, vol. 50, pp. 370–96.

MAYNARD, H. B., STEGMERTON, G. J., and SCHWAB, J. L. (1956), 'Methods–time measurement' in Maynard, H. B. (ed.), *Industrial Engineering Handbook*, McGraw-Hill.

MEETHAM, A. R., and HUDSON, R. A. (1969), *Encyclopaedia of Linguistics, Information and Control*, especially articles by Crossman, Gaines, Lange and Singleton, Pergamon.

MEISTER, D., and RABIDEAU, G. F. (1965), *Human Factors Evaluation in System Development*, Wiley.

MILLER, R. B. (1954), 'Psychological considerations in the design of training equipment'. *Wright Air Development Centre*, WADC TR 56-369.

MILLER, R. B. (1962), 'Task descriptions and analysis' in Gagné, R. M. (ed.), *Psychological Principles in System Development*, Holt, Rinehart & Winston.

MILLER, R. B. (1967), 'Task taxonomy: science or technology', in Singleton, W. T., Easterby, R., and Whitfield, D. (eds.), *The Human Operator in Complex Systems*, Taylor & Francis.

MORGAN, C. T., COOK, J. S., CHAPANIS, A., and LUND, M. W. (1963), *Human Engineering Guide to Equipment Design*, McGraw–Hill.

O'SHAUGHNESSY, J. (1966), *Business Organization*, Allen & Unwin.

POULTON, E. C. (1970), *Environment and Human Efficiency*, Thomas.

POWELL, P. I., HALE, M., MARTIN, J., and SIMON, M. (1971), '2000 accidents', Report No. 21, *National Institute of Industrial Psychology*, London.

PYM, D. (ed.), (1968), *Industrial Society*, Penguin.

ROGER, A. (1952), 'The seven-point plan', *National Institute of Industrial Psychology Paper No. 1*, National Institute of Industrial Psychology.

SEYMOUR, W. D. (1968), *Skills Analysis Training*, Pitman.

SHAW, A. G. (1952), *The Purpose and Practice of Motion Study*, Harlequin Press.

SHERIDAN, T. B. (1972) 'Supervisory control of teleoperators', in Bernotat, R. K., and Gartner, K. P., (eds.), *Displays and Context*, Swets & Zeitlinger.

SHERWOOD, S. L. (ed.) (1966), *The Nature of Psychology – Writings of the Late K. J. W. Craik*, Cambridge University Press.

SINGER, E. J., and RAMSDEN, J. (1969), *The Practical Approach to Skills' Analysis*, McGraw-Hill.

SINGLETON, W. T. (1957), 'An experimental investigation of sewing machine skill', *British Journal of Psychology*, vol. 48, pp. 127–32.

SINGLETON, W. T. (1959), 'The training of shoe machinists', *Ergonomics*, vol. 2, pp. 148–52.

SINGLETON, W. T. (1964), 'A preliminary study of a capstan lathe', *International Journal of Production Research*, vol. 3, pp. 213–25.

SINGLETON, W. T. (1966), 'Current trends towards systems design', *Ergonomics for Industry* No. 12, Ministry of Technology. Reprinted in *Applied Ergonomics* (1971), vol. 2, 150–58.

SINGLETON, W. T. (1967a), 'Ergonomics in systems design', *Ergonomics*, vol. 10, pp. 541–8.

SINGLETON, W. T. (1967b), 'The systems prototype and his design problems' in Singleton, W. T., Easterby, R., and Whitfield, D. (eds.) *The Human Operator in Complex Systems*, Taylor & Francis.

SINGLETON, W. T. (1968a), 'Some recent experiments on learning and their training implications', *Ergonomics*, vol. 11, pp. 53–9.

SINGLETON, W. T. (1968b), 'Acquisition of skill: the theory behind training design' in Robinson, J., and Barnes, N. (eds.), *New Media and Methods in Industrial Training*, BBC.

SINGLETON, W. T. (1969), 'Display design, principles and procedures', *Ergonomics*, vol. 12, pp. 519–31.

SINGLETON, W. T. (1970), 'Psychological aspects of manpower planning', A.P. Note 18, *Applied Psychology Department*, University of Aston in Birmingham.

SINGLETON, W. T. (1971a), 'Psychological aspects of man-machine systems' in Warr, P. B. (ed.), *Psychology at Work*, Penguin.

SINGLETON, W. T. (1971b), 'General theory of presentation of information', in Bernotat, R. K., and Gartner, K. P. (eds.), *Displays and Controls*, Swets & Zeitlinger.

SINGLETON, W. T. (1971c), 'The ergonomics of information presentation', *Applied Ergonomics*, vol. 2, pp. 213–20.

SINGLETON, W. T. (1972), *Introduction to Ergonomics*, World Health Organization.

SPENCER, J. (1965), 'Experiments on engineering drawing comprehension', *Ergonomics*, vol, 8, pp. 93–110.

SWAIN, A. O., and WOHL, T. G. (1961), 'Factors affecting the degree of automation in Test and Check-out equipment', *Dunlap and Associates* TR60–36F.

TUSTIN, A. (1947), 'The nature of the operator's response in manual control and its implications for controller design', *Journal of the Institute of Electrical Engineers*, vol. 94, pp, 190–202.

VANDENBERG, J. D. (1967), 'Improved operating procedures manuals', in Singleton, W. T., Easterby, R., and Whitfield, D. (eds.) *The Human Operator in Complex Systems*, Taylor and Francis.

WALLER, R. A. (1970), 'Making decisions about people and their physical environment' in Heald, G. (ed.), *Approaches to the Study of Organizational Behaviour*, Tavistock.

WALLIS, D. (1966), *Programmed Instruction in the British Armed Forces*, HMSO.

WARR, P. B. (ed.) (1971), *Psychology at Work*, Penguin.

WEBB, P. (ed.) (1962), *Bioastronautics Data Handbook*, NASA.

WELFORD, A. T. (1958), *Ageing and Human Skill*,
Oxford University Press, Reprinted in part in Legge, D. (ed.) (1971),
Skills, Penguin.

WHITE, W. J., and SCHNEYER, S. (1960), *Pocket Data for Human Factor Engineering*, Cornell Aeronautical Laboratory.

WHITFIELD, D. (1967), 'Human skill as a determinate of allocation of functions' in Singleton, W T., Easterby, R., and Whitfield, D. (eds.), *The Human Operator in Complex Systems*,
Taylor & Francis.

WULFECK, J. W., and ZEITLIN, L. (1962), Human capabilities and limitations', in R. M. Gagné, *Psychological Principles in System Developments*, Holt, Rinehart & Winston.

WULFF, J. J., and BERRY, P. C. (1962), 'Aids to job performance' in Gagné, R. M. (ed.), *Psychological Principles in System Development*, Holt, Rinehart & Winston.

Index